ENVIRONMENTAL SCIENCE, ENGINEERING AND TECHNOLOGY

AN INTERDISCIPLINARY APPROACH TOWARDS ACADEMIC EDUCATION ON SUSTAINABLE BUILDING DESIGN

ENVIRONMENTAL SCIENCE, ENGINEERING AND TECHNOLOGY

Additional books and e-books in this series can be found on Nova's website under the Series tab.

EDUCATION IN A COMPETITIVE AND GLOBALIZING WORLD

Additional books and e-books in this series can be found on Nova's website under the Series tab.

ENVIRONMENTAL SCIENCE, ENGINEERING AND TECHNOLOGY

AN INTERDISCIPLINARY APPROACH TOWARDS ACADEMIC EDUCATION ON SUSTAINABLE BUILDING DESIGN

VANJA SKALICKY KLEMENČIČ
VESNA ŽEGARAC LESKOVAR
AND
MAJA ŽIGART
EDITORS

NOTICE TO THE READER

The Publisher has taken reasonable care in the preparation of this book, but makes no expressed or implied warranty of any kind and assumes no responsibility for any errors or omissions. No liability is assumed for incidental or consequential damages in connection with or arising out of information contained in this book. The Publisher shall not be liable for any special, consequential, or exemplary damages resulting, in whole or in part, from the readers' use of, or reliance upon, this material. Any parts of this book based on government reports are so indicated and copyright is claimed for those parts to the extent applicable to compilations of such works.

Independent verification should be sought for any data, advice or recommendations contained in this book. In addition, no responsibility is assumed by the Publisher for any injury and/or damage to persons or property arising from any methods, products, instructions, ideas or otherwise contained in this publication.

This publication is designed to provide accurate and authoritative information with regard to the subject matter covered herein. It is sold with the clear understanding that the Publisher is not engaged in rendering legal or any other professional services. If legal or any other expert assistance is required, the services of a competent person should be sought. FROM A DECLARATION OF PARTICIPANTS JOINTLY ADOPTED BY A COMMITTEE OF THE AMERICAN BAR ASSOCIATION AND A COMMITTEE OF PUBLISHERS.

Additional color graphics may be available in the e-book version of this book.

Library of Congress Cataloging-in-Publication Data

ISBN: 978-1-53617-302-4

Published by Nova Science Publishers, Inc. † New York

The book title defines the didactic direction of the publication and the three basic contents: the interdisciplinary, the academic education and the sustainable building design, to be set in an interrelation. The title points to a complex undertaking for the formulation of problems, single professional aspects and guiding rules in the field of sustainable building design. The necessity of an interdisciplinary approach to the defined topic is argued very traceably in the introduction chapter…

The structure of the publication responds to the basic principle in the applied sciences: the definition of the complex problem and its aspects.

The different aspects (liveable neighbourhood, traffic environment, building energy modelling and structural design) are in single chapters and in a very comprehensible form and language. The interdisciplinary expert team of authors corresponds…offers a scientifically founded starting point for the design process.

The final part of the publication presents the development of the design ideas born in the frame of a workshop for a municipality in Slovenia…

The information offered in English is of interest not only for the recipient, the University of Maribor, but for students of architecture and also of a wider professional field including the academic field.

Grigor Doytchinov

Grigor Doytchinov, PhD
Associate Professor
Graz University of Technology,
Institute of Urbanism, Graz, Austria

Global challenges of sustainable development are strongly interlaced with the building sector. The changed relations towards the built environment, the resources and energy used in the building sector as well as the permanent investigations on the effects that the buildings have not only to the environment in general but to the health and well-being of their users and occupants require modifications in formation and education of designers and engineers who will be dealing with these emerging issues in the decades to come.

An interdisciplinary approach seems to be one of the key prerequisites for sustainable design. Current educational models, therefore, should be able to prepare future designers and engineers for productive and effective work in dynamic environments with permanent collaboration with various professionals and stakeholders. In the broad aspect of sustainable design practices and processes, the integration of disciplines can contribute to a more comprehensive design and a better understanding of the different aspects and challenges of other fields.

The monograph *Interdisciplinary Approach Towards Academic Education on Sustainable Building Design* represents a multidisciplinary collaboration between the architectural, urban, transport and construction professions to address various aspects of sustainable design in the educational process.

The monograph consists of a short introduction followed by six extensive chapters by different authors coming from different disciplines: an interdisciplinary approach in education, different aspects of creating liveable neighbourhoods, sustainable transport design and safety, integration of advanced digital technologies for energy performance simulations and structural design of sustainable buildings in timber construction. Together, they provide a systematic approach to the educational process in sustainable building design to ensure environmental protection and quality of life. The sixth chapter is essentially a graphical appendix with the presentation of the final results of the student design workshop, in which the implementation of the different aspects and themes has occurred.

The monograph is clearly structured and written in a way that can serve as a tool for various experts and professors in the field of sustainable building design education. The topics presented cover some of the key aspects of sustainability in contemporary building design. Through individual chapters, a strong integration of scientific research findings into the educational process is visible, as individual research topics emerge as guidelines for teaching and effective collaboration through interdisciplinarity.

The idea of an interdisciplinary approach is no longer new in the educational process, but it is introduced in the book in a manner that emphasizes both interdisciplinarity in student groups and interdisciplinarity within mentors and consultants, and highlights extensive expert linking with the aim of creating a dynamic learning environment. A positive contribution of the presented workshop is the demonstration of the effective and fruitful cooperation between the University of Maribor, Faculty of Civil Engineering, Transportation Engineering and Architecture (UM FGPA) with the Municipality of Podlehnik in Southeast Slovenia and the involvement of local communities.

Nataša Ćuković Ignjatović, PhD
Assistant Professor
University of Belgrade, Faculty of Architecture,
Department of Architectural Technologies, Belgrade, Serbia

The book entitled, *Interdisciplinary Approach Towards Academic Education on Sustainable Building Design*, consists of contributions written by various scholars including Vesna Žegarac Leskovar, Vanja Skalicky Klemenčić, Marko Renčelj, Maja Žigart and Miroslav Premrov, from the Faculty of Civil Engineering, Transportation Engineering and Architecture of the University of Maribor (UM FGPA), Slovenia.

The topic of the book is very complex and comprehensive, since, as the title implies, it covers three broad and very significant topics which multiply their significance when correlated. Although the significance of education on sustainable development was already recognized in the 1970s, it is still a very current topic in scientific and teaching circles, and has become a new field of study and research.

The intention of the publication is to present the importance of an interdisciplinary approach towards academic education on sustainable building design, which is systematically presented through the topics covered in the book chapters. The publication contains theoretical knowledge on sustainability at different planning and design levels with a prominent scientific approach, as well as an overview of their practical applications.

The first chapter indicates the importance of an integrative approach in education on a sustainable built environment, with an emphasis on the educational process with a multilevel interdisciplinary approach implemented at the UM FGPA in Slovenia. The following chapters cover various topics on sustainable building design (energy-efficient buildings, timber construction for low environmental impact, liveable environments and sustainable traffic design to achieve human well-being) that are relevant to achieve a broad range of knowledge in the subject area for further planning and design processes. Accordingly, the final chapter elaborates the application of acquired knowledge from various aspects of sustainable building design…

All chapters have a scientific character and are provided with appropriate figures, tables and references. From the topics covered, it is clear that this publication wishes to emphasize that the living environment should be planned in a holistic manner, relating the building to the environment and vice versa. Despite the different research topics presented, it is possible to detect challenges and goals common to all disciplines and consequently addressed from different perspectives.

This book can be used in university teachings and which could also serve as an incentive for academics to think about starting new courses based on different forms of teaching while engaging experts from different scientific fields.

Based on the above-mentioned conclusions, I recommend this scientific monograph for publication with the conviction that it will contribute to understanding the comprehensiveness of education and sustainable building design topics, and will contribute to promoting awareness of the importance of interdisciplinarity and a holistic approach in the education of future architects and engineers.

Lea Petrović Krajnik, PhD
Assistant Professor
University of Zagreb, Faculty of Architecture,
Department of Urban Planning,
Spatial Planning and Landscape Architecture,
Zagreb, Croatia

CONTENTS

PREFACE

The significance of education on sustainable development was first recognized in the early 1970s by UN leaders at the United Nations Conference on the Human Environment. This first major conference on international environmental issues, which was held in Stockholm in 1972, was a turning point in the development of international environmental politics. The UN leaders adopted the Stockholm Declaration that revealed the primacy of education on environmental matters as a means toprotect the environment in its full human dimension. Furthermore, the declaration called attention to scientific research and development in the context of environmental problems to be promoted in all countries. With increasing attention to climate change affecting the gradual deterioration of the environment, education on sustainable development has become a new field of study and research over the last two decades. Especially in engineering and architectural academic institutions, the integration of sustainability issues into regular educational curricula opens new challenges and approaches to the teaching process. The consideration of sustainable design aims to widen the students' perception of the global environmental situation, and establish an attitude towards social responsibility, while encouraging innovative thinking and problem solving. However, education on sustainability can be identified as a very complex field of study which requires the understanding of various perspectives. Therefore, it is often taught within the interdisciplinary context which encourages the collaborative approach.

This book was written to present the interdisciplinary approach to academic education on sustainable building design on the case of the University of Maribor, Faculty of Civil Engineering, Transportation Engineering and Architecture (UM FGPA). It consists of six chapters. The first chapter describes the general aspect of the integrative approach to education on sustainable built environment through the experience of the interdisciplinary course "Sustainable aspects of building design" carried out at UM FGPA. The emphasis is put on the presentation of the educational process, in which the multilevel interdisciplinary approach was implemented to enable students to obtain comprehensive knowledge. Disciplines like urban design, sustainable transportation, energy-efficient building design, and structural design of timber buildings were introduced during the course. This chapter presents the effectiveness of such education and exposes some weak points that have to be upgraded to achieve the desired quality. Chapter 2 introduces a comprehensive set of urban design criteria for creating residential neighborhoods as a tool for urbanist, architect and policy makers. The main objective of the chapter is to define a system of criteria with the holistic approach for high-quality residential environments – not just the building design but the design of a residential environment as a whole. The criteria are discussed in detail and implicated within the pedagogical process at UM FGPA. Chapter 2 emphasises the role of high-quality open space and green areas, which significantly contributes to liveability–human-centered urban design. Chapter 3 deals with sustainable transport in connection with the living environment, and emphasises the interdisciplinary approach in the field of higher education in the field of sustainable buildings design. Being aware of the negative consequences of traffic, especially road traffic, sustainable mobility is an established approach that can reduce the negative effects of road traffic. In the field of the living environment in connection with mobility, this chapter also describes key steps: sustainable mobility, sustainable traffic safety, (road) and infrastructure measures to ensure sustainable mobility. The final section presents examples of an interdisciplinary search for solutions at the level of student tasks and master's theses. Chapter 4 presents the inclusion of energy-efficient building design with the integrated use of new digital technologies. It shows how the sustainable design process can be applied in the education process within interdisciplinary courses with an emphasis on achieving living comfort, energy efficiency, and low environmental impacts in buildings by implementing building information and energy modelling. The first part of Chapter 5 briefly describes the main types of timber structural systems, while the second part presents a sustainable design perspective of contemporary prefabricated timber-glass buildings. Due to the fact that contemporary design of timber-glass buildings can cause many structural problems, special prefabricated timber-glass wall elements were developed, which is also shown in this chapter. Many previous studies and international research projects are briefly described at the end of the chapter, which can open many other opportunities in further development of multi-story prefabricated timber buildings with enlarged glazing areas. The last chapter is prepared as a graphic summary of interdisciplinary workshop projects which demonstrate the complexity of design and the respect for diverse interdisciplinary principles of sustainable planning.

In: An Interdisciplinary Approach … ISBN: 978-1-53617-302-4
Editors: V. S. Klemenčič et al.

Chapter 1

INTERDISCIPLINARITY IN ACADEMIC EDUCATION ON SUSTAINABLE BUILT ENVIRONMENT

Vesna Žegarac Leskovar*, PhD
University of Maribor, Faculty of Civil Engineering,
Transportation Engineering and Architecture, Maribor, Slovenia

ABSTRACT

The built environment has been under worldwide scrutiny regarding building quality and performance through intensive consideration of its environmental impacts on the one hand, and the influence on public health and the well-being of its users on the other. Respecting both the aforementioned priorities, the trends in current building design tend to be in line with multiple dimensions of sustainability. In addition to its occurrence in scientific research and practice, the topic of sustainable design is ever more emerging within the curricula at universities, whereby educational approaches can be very effective if interrelated to different scientific or expert disciplines.

---

* Corresponding Author's Email: vesna.zegarac@um.si.

This chapter presents the integrative approach to education about sustainable built environment through the experience of the interdisciplinary course "Sustainable aspects of building design" carried out at the University of Maribor, Faculty of Civil Engineering, Transportation Engineering and Architecture. The course is part of master programmes of architecture and civil engineering and is carried out in a form of an elective subject, which has been progressively developing. The main challenge of the course is to implement a multilevel interdisciplinary approach. The aim of this chapter is to present the effectiveness of such education and also to expose some weak points that have to be upgraded to reach the desired quality. Nevertheless, the presented education approach strives for comprehensiveness with regard to several integrated disciplines and is a valuable contribution to sustainable development.

Keywords: academic education, interdisciplinarity, sustainability, built environment

INTRODUCTION

Witnessing climate change and its related consequences, raising awareness on sustainability has become a key issue of various policies, economy sectors, and educational institutions. The need for a more sustainable environment has also been recognized by world leaders at the United Nations Summit in September 2015, when 17 Sustainable Development Goals (SDGs) of the 2030 Agenda for Sustainable Development were adopted (UN, 2015). The strategy of the agenda indicated the importance of high-quality education as a factor that greatly contributes to the creation of sustainable development. As sustainability challenges can no longer be ignored, the topic of sustainability has been introduced in many engineering and humanistic higher education curricula over the world recently, whereby the main aim is to establish innovative ideas and approaches regarding how to combine sustainability and disciplinary knowledge, skills, and experiences in the education scheme (Cincera et al. 2018). The process of integrating sustainable development education in academic institutions is very complex, and requires either a redesign of the existing programmes and courses or a creation of new ones.

Not only the content of the courses but, also new approaches to teaching and learning have to be adopted (Cincera et al. 2018, Vila et al. 2012). Advancing the academic curricula requires the integration of comprehensiveness and holism in knowledge sharing to create solution-oriented study programmes (Tejedor et al. 2018). To develop the sustainability competencies of students, the collaborative, interactive, interdisciplinary and value-based teaching and learning is recommended (Christie et al. 2013). According to (Cincera et al. 2018) also the competencies of teachers, researchers, and practitioners should be advanced in the process of implementing education on sustainable development into the higher education environment. One of the global ideas of such education is to encourage a critical and creative thinking processes with a focus on social equity and empathy, which opens up opportunities for society to develop in the near and far future within current ecological, socioeconomic, and political challenges. As agreed by most academics, education should not be focused merely on the acquisition of knowledge, but on the development of the ability to apply that knowledge to solving practical problems. In this manner, education on sustainability involves discussions on serious environmental and social issues in the local and global context, but also practical solutions that could be applied by integrating real-world situations into teaching and learning (Douvlou 2006).

Interdisciplinarity in Education on Sustainability

The abovementioned challenges, and the increasing global need to increase efficiency and quality of projects over their whole lifecycle have given rise to the demand for new approaches in education on architecture, engineering, and construction. Studies have been conducted recently, which analyse the abilities required from young professionals in urban planning, architecture, engineering, and the construction industry (Becerik-Gerber et al. 2011, Li et al. 2018), who are responsible for the quality of the built environment. These studies indicate that, apart from the fundamental professional knowledge and computing competencies, very important skills are collaboration and teamwork, in addition to having a wider perspective on social, environmental, and economic issues related to a certain profession (Becerik-Gerber et al. 2011). Due to the current rapid technological progress, it is of utmost importance for young professionals to constantly update their knowledge and to be able to apply it directly in practice. The research in (Tejedor et al. 2018) even indicates that most interdisciplinary initiatives in technological education for sustainability fit into a problem-solving discourse, where the co-production of knowledge and method-driven aspects are relevant. Becerik-Gerber et al. (2011) also argue that the important challenge of the academic education of young professionals is to maintain the connection to real-world practices which address present and future challenges of the built environment industry from the aspect of sustainability. In their study, they examined higher education programmes in the USA related to architecture, engineering, and construction. The results of the analysis conducted within the study show that most programmes perceive sustainability as very important for the future of the architecture, construction, and engineering industry. The sustainability topics mostly considered are sustainable design, sustainable construction practice, environmental analysis, energy conservation, renewable energy, materials, and green building certification systems. However, particular programmes give different priority to these topics with regard to a specific profession. On the other hand, many studies emphasise that architecture, engineering, and construction education, which are responsible for the competencies of future experts who will create our built environment, need to change focus from a single discipline to the interdisciplinary approach (Becerik-Gerber et al. 2011, Cincera et al. 2018, Tejedor et al. 2018, de Gaulmyn and Dupre 2019). In higher education, the term "interdisciplinary" describes programmes that combine or involve knowledge of, and insights into, two or more established academic disciplines or fields of study or professions. The interdisciplinary approach in education may increase the ability to understand complex challenges, facilitates problem solving, and promotes better understanding of problems (Eagan et al. 2002). According to (Clark and Wallace 2015), disciplinarity related to the treatment of individual disciplines may be further classified. They argue that knowledge may be classified as disciplinary (when disciplines work in isolation), multidisciplinary (when disciplines work in parallel to address common goals), interdisciplinary (when disciplines work in an integrated way), and transdisciplinary (when there are no limits between disciplines).

However, interdisciplinary education can be implemented in various ways. It either enables students from different disciplines to join the same course and collaborate in a team environment, or teachers who create a course come from different disciplines or even people who work outside academia can conduct or supervise a part of the course. On this basis, it is obvious that interdisciplinarity can be attained not only among individual disciplines which construct a common model (Tejedor et al. 2018), but also between the academic environment and practice. Although the interdisciplinary approach in architecture and other engineering sciences which address sustainable built environment is needed due to highly complex current challenges that require a holistic view, students are not expected to become experts in all disciplines, but rather to widen their insights and to obtain the skills to work with experts from different fields in a team. The study of Segalàs et al. (2010) even emphasises that engineering students, in addition to the basic competency of systemic thinking, should also have the competency of interdisciplinarity upon graduating. When integrating sustainability in architecture, construction, and other engineering sciences, it is unavoidable to combine ecological, technical, social, and economy perspectives to develop interdisciplinary knowledge, tools, and approaches, which will help society develop solutions (Annan-Diab and Molinari 2017).

This chapter introduces the integration of the interdisciplinary approach to education on the case of the University of Maribor, Faculty of Civil Engineering, Transportation Engineering and Architecture (UM FGPA) through the implementation of various activities related to the development of innovative approaches towards sustainable design of the built environment. As universities are recognized as complex organizations, which inevitably evolve over the long term in response to changes in the societies (Bozgut 2006), the process of integrating interdisciplinarity is very broad in scope. In addition to basic educational curricula, it combines the implementation of practical student workshops, development projects, scientific research, and various activities aimed at raising awareness and the spreading of knowledge. The contemporary educational process is predominately conducted in the context of current environmental issues and also in terms of specific problems which arise from the local environment and economy. Offering four separate study programmes, i.e., Civil Engineering, Transportation Engineering, Industrial Engineering, and Architecture, the faculty has a great potential for the integration of different disciplines in the basic curricula, where specific elective courses can be adjusted to the requirements of a particular study programme. Within the educational process, interdisciplinarity is met on multiple levels (Figure 1).

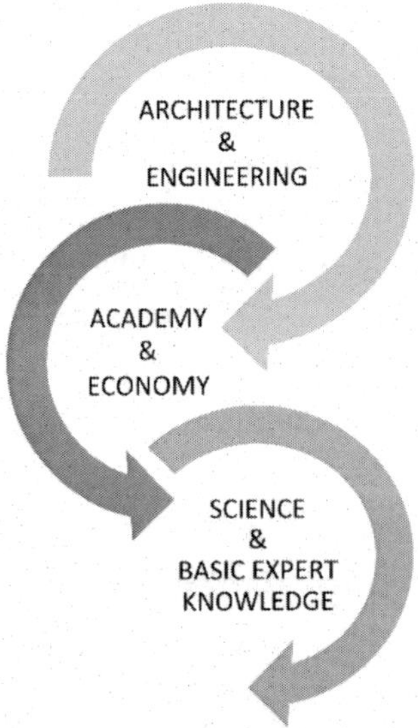

Figure 1. The principle of interdisciplinarity at UM FGPA.

On one of the first levels, the implementation of the latest scientific findings into the theoretical and practical educational courses enables education to be in line with the actual design trends. Secondly, the interdisciplinarity of different expert fields, such as urban, architectural and structural engineering, energy efficiency, the quality of indoor environment, and environmental engineering in accordance with sustainable design aspects, enables the educational process to gain comprehensiveness with regard to multiple integrated disciplines and approaches. Finally, interdisciplinarity is indicated through the collaboration of academic circles with industry and the local environment. Integrating interdisciplinary approach into education at UM FGPA lays foundations for a path towards the realisation of an important goal of contemporary education, which is to enable the students to gain fundamental expert knowledge upgraded with competencies of problem solving on the local and global scale, which can contribute to raising the quality of our built environment in the future.

INTERDISCIPLINARY APPROACH TO SUSTAINABLE DESIGN EDUCATION AT UM FGPA

Education holds a valuable role in complementing knowledge with past experiences (Bozgut 2006), which, however, should be upgraded with novelties which reflect present and expected development. The main idea underlying the education scheme at UM FGPA is the connective interrelation of individual main activities which constitute a comprehensive circular scheme (Figure 2).

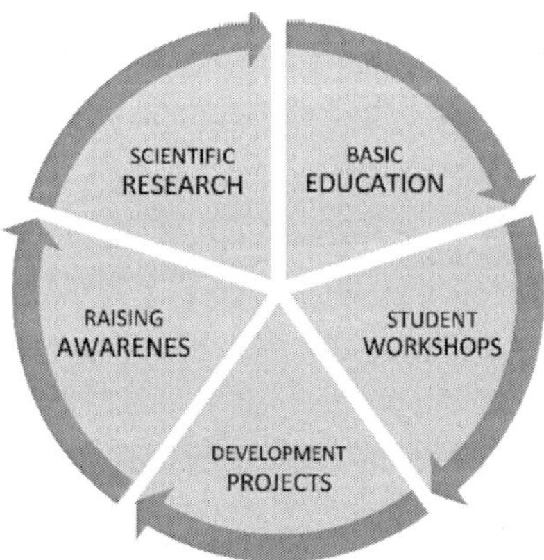

Figure 2. Interrelation of activities meaningful for the provision of high-quality education.

Table 1. Levels of interdisciplinary education in academic education at UM FGPA

	Traditional courses	**Interdisciplinary courses**		
		Level 1	**Level 2**	**Level 3**
Form of the course	Core	Core or elective	Core or elective	Core or elective workshop selected students
Students	From one discipline	From one discipline	From two or more disciplines	From two or more disciplines
Students activity	Passive	Active	Active - decision making	Active - investigating - decision making
Practical work (project)	Not necessarily	Yes	Yes	Yes
Students collaboration	Individual work	Work in pairs	Work in interdisciplinary pairs or teams	Work in interdisciplinary teams
Teachers from academia	From one discipline	From two disciplines	From two or more disciplines	From two or more disciplines
Experts outside academia	/	/	Sometimes involved	Involved
Other stakeholders	/	/	/	Involved

In addition to knowledge based on facts and past experiences, the integration of scientific research findings into educational curricula, especially into master courses, provides students with direct information on novelties and keeps their perception up-to-date. Especially in introducing and teaching the topic of sustainability, which is very complex in terms of its correlating perspectives, the educational process requires drastic changes to conventional pedagogical structures and methods (Li et al. 2018). In this manner, certain courses concerning sustainability are transformed into interdisciplinary courses in terms of content and teaching methods. These courses still generally consist of the theoretical and practical part, while they differ from traditional disciplinary ones in the implementation methods. In terms of disciplinarity, the courses at UM FGPA can be classified into three levels (Table 1).

As an upgrade of traditional courses, Level 1 includes a soft interdisciplinary approach in which students play an active role in the learning process by contributing to smaller projects and debates, which encourages critical thinking. While students still come from the same discipline, collaboration between educators contributes to expanding the knowledge beyond borders of one expert or scientific fields. Level 2 is the intermediate level of interdisciplinarity. Students from two or more disciplines work actively on problem solving in pairs or groups. These courses are simultaneously carried out for students of different bachelor study programmes, for example students of architecture and civil engineering, and conducted by teachers of different disciplines to encourage interdisciplinary collaboration in the early education stages. The courses of master curricula are far more specialised; therefore, they are conducted separately for different study programmes in the first, i.e., theoretical, phase. While the theoretical part is adjusted to students of different disciplines, joint work is applied in the practical part of the courses in which students have to create a design. In the practical part of the course, students of different disciplines usually meet on problem-solving projects. Problem solving and design thinking skills are often identified as a key outcome of the architectural studio pedagogy (Gaulmyn and Dupre 2019). Usually, students of architecture create preliminary design proposals as conceptual responses to a design problem. In disciplinary architecture courses, students commonly move back and forth between ideas or theories to develop multiple proposals before determining the most suitable solution. The process is quite similar in interdisciplinary courses. The difference to pure architectural courses is in the scope and perspectives of empirical and theoretical considerations that stimulate students to put forward their best proposals, since this process also includes students of civil and industrial engineering, who integrate their knowledge in order to support and achieve the comprehensiveness of design projects. It is really important in such an educational process to have good communication between students and professors from different expert fields. In this kind of courses, it is possible that also experts from the non-academic environment supervise and support part of the theoretical or practical work. Finally, the most intensive level of interdisciplinarity is integrated into special workshops that are often conducted outside the frames of regular courses. For better insight into the real practice problems, the courses can be partly carried out in the form of student workshops in collaboration with parties representing the local environment and businesses. In addition to academic supervisors, this kind of active education through integration of external experts includes expert consultants from the field of architecture, urban planning, construction, transport, and other engineering sciences, whereby the whole planning process highly resembles the processes of real project planning and design development (Figure 3). This level is largely conducted by selected advanced students who show a high interest in certain projects.

Source: Archive UM FGPA.

Figure 3. Students of architecture and civil engineering consulting with the expert from a mechanical engineering company for the purposes of an interdisciplinary student workshop.

Focusing on the development of new innovative building design solutions, such projects are predominately requested by businesses or local municipalities. The outcomes are important not only for students who gain new experiences, but also for stakeholders who gain innovative ideas. Interdisciplinary projects strengthen the bonds between the academy and the economy, which is particularly significant for overall economic competitiveness and progress. Additionally, the implementation of common development projects brings first employment opportunities to students involved, while partners from the business sector have the possibility to meet potential employees.

Source: Archive UM FGPA.

Figure 4. Exhibition "Lessons on timber construction" at the Chamber of Commerce and Industry of Slovenia, which presents various development projects of UM FGPA carried out in interdisciplinary workshops.

Last but not least, the gained knowledge and experiences presented by the comprehensive circular activities scheme (Figure 2) may not be retained within the boundaries of universities and partner organisations involved. A mission of universities is also to contribute to the welfare of society, as the results of systematic activities are shared with public. Awareness is raised through the publishing of the results in local magazines, national and international professional and scientific journals and monographs, and through active presentation to the wider public at exhibitions, symposiums and conferences organised by a university (Figure 4).

The described activities are planned as an integral process aimed at the on-going optimisation of main findings and subsequent interactions within this specific circular education scheme.

CASE STUDY - INTERDISCIPLINARY STUDENT WORKSHOP

This subchapter presents a comprehensive circular education scheme which demonstrates the process of the interdisciplinary education approach. The case study presented in this subsection refers to the educational process conducted in the 2017/2018 winter semester within the course "Sustainable concepts of building design" at UM FGPA. The course is offered in the master study of Architecture and Civil Engineering, and has been developed to integrate students from various programmes, who already have fundamental knowledge of their primary disciplines, to approach problem solving in interdisciplinary teams. The aim of the course is to upgrade basic knowledge of students on urban design, architecture, and construction in the light of sustainable development goals. Students are guided to examine design challenges from social, ecological, and economic perspectives, define and analyse problems, explore the value indicators of sustainable built environment, define aims and objectives, and to develop and verify their proposals for problem solving using socially responsible design thinking. The course basically is conducted and supervised by professors and assistants in architecture and civil engineering, while also teachers from other fields can be involved, which depends on the design project. This collaborative structure is constantly adapted and improved throughout the process, allowing students from various academic disciplines to contribute to their projects, while learning and benefiting from the perspectives of other disciplines (Li et al. 2018). In order to come closer to real problems regarding the environment or the economy, the topics considered in the course are usually selected in collaboration with external stakeholders, e.g., companies, municipalities, ministries or other organisations. Such a process encourages students to adapt to different mindsets, and to work with different stakeholders with diverse technical, social or cultural backgrounds, while understanding and reacting to conflicts of interests that may arise among different stakeholders.

General Characteristics of the Area Considered by the Project

The design project entitled Punkt Podlehnik (in Slovenian slang "punkt" means "point") selected for the course in the 2017/18 winter semester addressed the challenge of sustainable development of the Municipality of Podlehnik. The rural area of Podlehnik is located in the north-eastern part of Slovenia (Figure 5) and covers 46.0 m^2 of land that is very diverse and characterised by hills full of vineyards (Figure 6).

Source: Archive UM FGPA.

Figure 5. The Municipality of Podlehnik is located in the north-eastern part of Slovenia.

Recently, the new Draženci–Gruškovje motorway which crosses the Municipality of Podlehnik has been constructed (Figure 6). The newly constructed segment is part of the motorway connection between international border crossings with Austria and Croatia. Therefore, it unlocks the potential for an increase in international traffic through the area, and makes Podlehnik more connected to other Slovenian and neighbouring cities. The demographic structure of the municipality with app. 1,800 inhabitants shows that most inhabitants are between 30 and 69 years old, while an evident lack of the younger generation of up to 29 years of age (Figure 7) is one of the key problems of the ageing municipality. This problem was also one of the main causes for the authorities of the municipality inviting UM FGPA to participate in a design workshop. Their ambition was to obtain new ideas that would make the municipality more attractive to younger generations and to tourists passing through the area on the newly constructed motorway, while upgrading the quality of the environment for ageing local inhabitants. More specifically, the authorities addressed two areas where they expected major spatial interventions. The first area is located in the very centre of the Municipality of Podlehnik near the motorway, where, according to the project task, residential buildings, a hotel facility, and an intergenerational centre were envisaged. The second area is an idyllic location near the Dežno Pond, where the desire of locals was primarily aimed at the development of a new tourist area.

Figure 6. A view of the Podlehnik area (Photo by: Marko Vindiš).

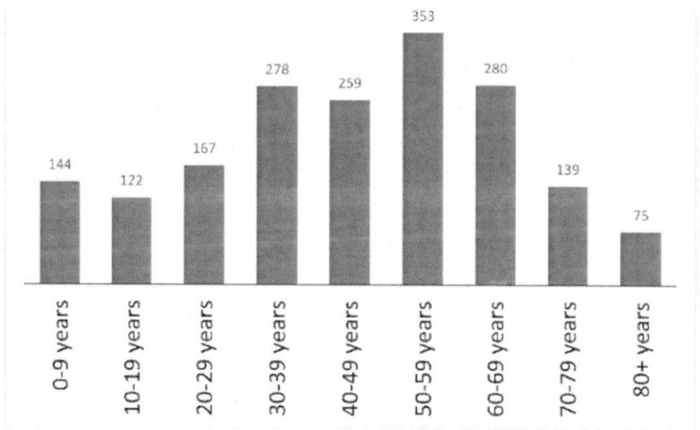

Figure 7. Age structure of the Municipality of Podlehnik (Adapted from: https://wwwtypopulation/en/slovenia/admin/podravska/172__podlehnik/, accessed 12 June 2019).

Model of the Workshop

In accordance with the basic information and wishes received by the city council, the task for the students of architecture and civil engineering who attended the workshop was to analyse the characteristics and potentials of the area to develop proposals and solutions that would contribute to sustainable development of the municipality. Due to its highly interdisciplinary nature, the workshop was conducted by professors and consultants from different chairs of UM FGPA, i.e., the Chair of Architecture, the Chair of Spatial Planning, the Chair of Structural Engineering, and the Chair of Traffic Infrastructure, while certain phases of the workshop also included representatives of the local community. The theoretical knowledge shared by faculty professors consisted of the basic theory, and of their own and general latest scientific findings, which influenced the decision-making process of students generated in the practical phase of the course. The scheme of the workshop is shown in Figure 8. The smaller dark dots mark the activities held in the university environment, while the bigger grey circles denote the activities implemented in the environment of Podlehnik.

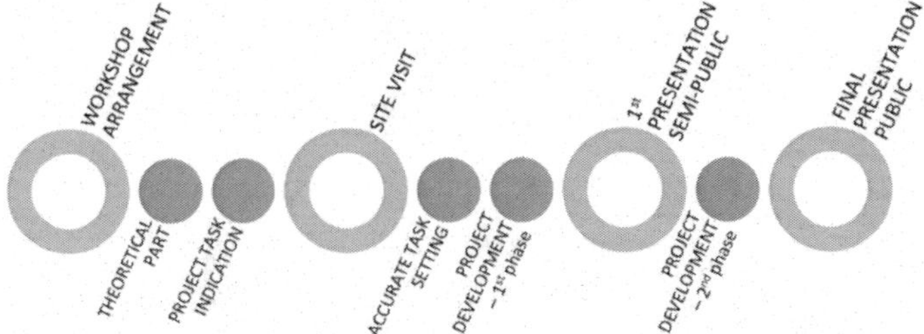

Figure 8. Scheme of the Punkt Podlehnik workshop.

(a)

(b)

Source: Archive UM FGPA.

Figure 9. (a) Meeting and discussion with the authorities of the Municipality of Podlehnik, and (b) site visit.

The process began with the first meeting of university professors and municipality authorities (the mayor, municipal councillors, the technical director) in the Podlehnik area to set the framework and define the main challenges of the workshop. The students joined the project in its theoretical part, in which the lectures on sustainable design fragmented into different topics, from the basics of sustainability perspectives to more specified themes, such as sustainable spatial planning, green architecture, energy efficiency, lifecycle environmental impact analysis, sustainable construction, timber construction, transportation planning, etc., were delivered by professors of various disciplines. The lectures were continuously followed by discussions with students expressing their opinions, understanding, thoughts, and arguments. After the theoretical part, the professors presented the assignment of Punkt Podlehnik to the students in order to share some basic information on developmental challenges of the observed area. To understand, discuss and obtain a critical opinion on these challenges, it was necessary for the students to interact with the community. Therefore, the next step of the workshop was a site visit and a meeting with representatives of the Municipality of Podlehnik (Figure 9).

As the assignment needed to be understood in the context of the natural and built environment, the built infrastructure, the economy, social status, and first and foremost, human communities, a closer contact with the area under consideration was necessary for the students to be directly confronted with the real-case situation. In this manner, the workshop continued as an experimental learning process in which the development challenges discussed in the course began to be contemplated within local, regional and global contexts. Not only the site visit and discussion with municipality authorities, but also contact with local residents encouraged the students to identify distinct challenges of urban development, explore the historical context and contemporary urbanisation trends, investigate likely new contents that would attract immigration of younger generations to the municipality, and to review the tourist potential. The next step of students design process was the accurate task setting, in which they defined project goals and perspectives to start the first stage of project development. Under the supervision of university professors and assistants from the fields of spatial planning, architecture, transportation engineering, and construction, the students joined interdisciplinary teams to initiate accurate analyses of the site and start evolving their first ideas, while tackling the challenges of balancing cultural heritage, land use and water resources, environmental quality, economic restraints, and social values. One half of the students addressed the centre of Podlehnik and the other half approached the development of a tourist area of the Dežno Pond. The first development stage lasted for about four weeks, and regular meetings with academic supervisors were held once a week.

The teams addressing the Dežno Pond area explored the possibilities for developing primarily tourist activities by incorporating new attractive programmes and setting up temporary and permanent tourist modular units made of the timber-glass structure. After a detailed site analysis, they found that the pond was poorly visited, as it does not provide sufficient activities attractive to tourists. However, the students pointed out the potential the pond area had for the development of recreational and relaxation activities. They anticipated that a diverse programme would attract tourist of different age groups to visit the area throughout the year (Figure 10). The student teams came up with different concepts for temporary and permanent accommodation, from the increasingly popular glamping, smaller modular timber prefabricated modular houses, and open space houses to multi-bed units for larger school groups. In their designs, students introduced interventions not harmful for the original natural environment. In this context, they proposed the design of prefabricated timber-glass modular units, the reuse of existing facilities and sports fields, as well as landscape arrangements that take advantage of the natural features of the pond and its surroundings.

The teams which developed proposals for the centre of the Municipality of Podlehnik pointed out the advantages of the Podlehnik location, such as its connectivity to other cities and easy access to jobs, exceptional natural resources of the municipality, and potentials for the development of transit and rural tourism (Figure 11). In the development of the central area with predominantly residential use and additional public programmes, it was important to limit noise and pollution that could be caused by the proximity of the new motorway, and to anticipate activities on the site that would revive the existing centre of the Municipality of Podlehnik providing useful and attractive programmes to future residents.

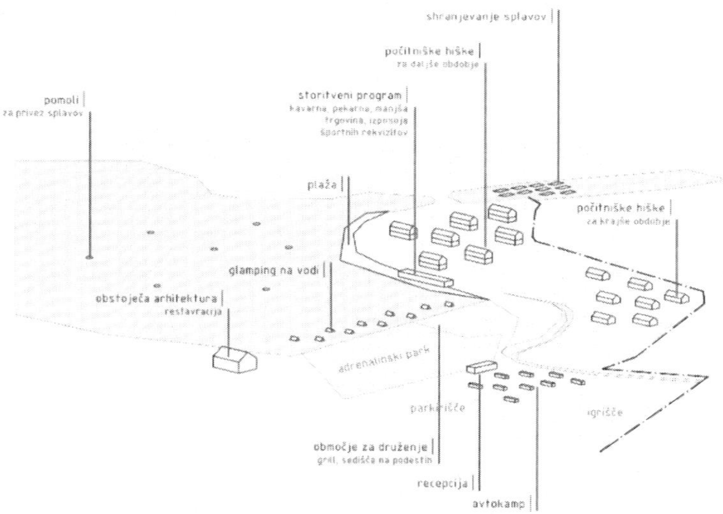

Source: Archive UM FGPA; design by students of UM FGPA - Špindler S., Vuković I.

Figure 10. First design concepts for the development of a tourist area along the Dežno Pond.

Source: Archive UM FGPA; design by students of UM FGPA – Dragovič J., Nežmah K., Rogač S., Sekereš S.

Figure 11. Analysis of the centre of Podlehnik.

First design concepts were based on the theoretical knowledge gained in the first part of the course, references to similar projects, and on findings gathered from site analyses, where the communication to community representatives played an important aspect. All teams that dealt with the central area incorporated the residential programme on the east side of the site, exposed to a quieter, hilly green vineyard slopes, and envisaged a mixed-use hotel, intergenerational centre intended to host older inhabitants, and other mixed-use public facilities along the existing local road to the west. These facilities would thus be a dividing line between the noisy highway in the west and the quiet housing programme in the east, and continue the formation of the public central area with the main square. The students put a lot of effort in developing a programme with a balanced interrelation between the built-up area and open space. Apart from the guidance related to urban design, a supervisor from the transportation planning department instructed students on appropriate design solutions for transport network, delivery and intervention routes (Figure 12).

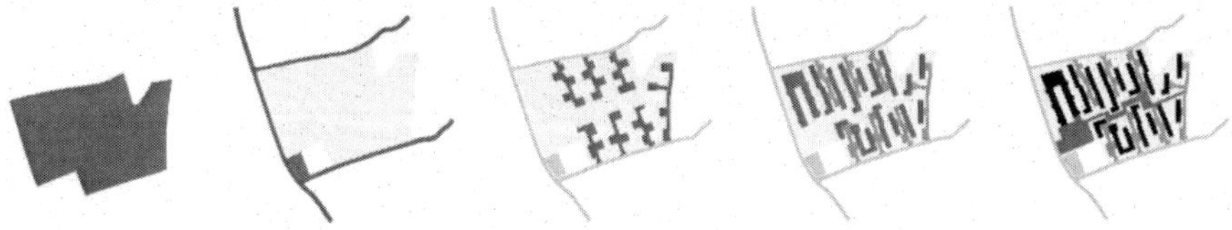

Source: Archive UM FGPA; design by students of UM FGPA – Belec T., Božak T., Leskovar D. Tolazzi G.

Figure 12. First design concepts for the central area of the Municipality of Podlehnik.

When developing a residential programme, the students incorporated sustainability perspectives into their design proposals for contemporary individual and multi-family housing typologies, aiming to spark interest of younger generations and young families to settle in the Podlehnik central area. While the housing facilities were designed as flexible units that can be adapted to individual needs, the supplementary semi-private common spaces were incorporated in residential buildings to encourage community generation, work from home, and a high-quality social life. The rest of the programme complemented the already existing public programme of the municipal facility with a health-care centre, an elementary school, a kindergarten, a shop, and a police station located in the centre of Podlehnik. In this manner, the students created new supplementary public and semi-public programme with a hotel for transit tourists, an intergenerational centre with housing options for older people, and accompanying service activities. The students also planned diverse public and semi-public areas, smaller markets and squares, green areas, and outdoor sport facilities, such as a skate park and an outdoor fitness area in conjunction with the existing elementary school.

The built environment under discussion was arranged in the light of sustainable design concepts which take into account ecology viewpoints, energy efficiency principles of passive and active design strategies, the economical use of space and materials, market oriented and human-centred design, all aiming to achieve a high quality of life. BIM software tool Archicad Graphisoft was used for the planning process, so the energy performance was ensured simultaneously to the development process of new design solutions. Furthermore, daylight analyses were carried out with a Velux Daylight Visualizer. When defining building materials, the students put emphasis on ecological values and lifecycle environmental performance; therefore, the prevailing building material used in the project was timber. Respecting the structural demands of newly designed building volumes, students of civil engineering collaborated already in the early design stage to support the architecture students in their design decisions. Furthermore, they also conducted structural analyses of all newly designed buildings after the preliminary designs had been finished.

The first development stage of the workshop showed the importance of interdisciplinary cooperation between different departments, and interaction among students of architecture and civil engineering. To discuss the project proposals with representatives of the community, the first semi-public presentations were carried out in the Municipality of Podlehnik. The mayor, the vice mayor, and municipality counsellors attended the presentations and provided the students with feedback to the project in the form of critical opinions, remarks, and open questions. The first presentation stage was very important for the project development, as the students checked the relevance of their task setting, and also had the opportunity to defend certain ideas that were, to a certain extent, misunderstood or unusual for community representatives. By involving stakeholders from the local environment in the workshop, the interdisciplinarity intensity rose to an even higher level. Not only the academic staff, but also people from outside academia had the opportunity to take part and influence the project development.

Source: Archive UM FGPA.

Figure 13. Public presentation of the student project at the main hall of the Podlehnik municipality building.

After initial semi-public presentations, the workshop progressed into the second development stage in which, based on the discussions from presentations and the re-evaluation of the attainment of project goals, the task setting stage was defined and the students upgraded their initial design proposals. After approximately four weeks of additional work in interdisciplinary teams, the students completed their projects and prepared graphic materials to be exhibited and presented to the public at an exhibition. The final presentation in front of a panel of public audience composed of the city council, experts, inhabitants and potential clients who were interested in investing in the project was held in the public hall of Podlehnik municipality building at the end of the winter semester in February 2018 (Figure 13). During the final presentation, each team pointed out the main concept and details related to their project proposals. After the public presentation, the teams were invited to actively join the exhibition of their works, where they could additionally explain, disseminate, and argue their ideas to potential investors who gave direct feedback to the students. The dissemination of the project continued after the public presentation through posts in social media, and the publication of articles in local and national magazines.

The whole educational process was very complex with constant interaction between the academic and local environment, enabling the students to learn from a real-life situation and to put the educational process into practice.

Main Outcomes of the Interdisciplinary Process

The course “Sustainable concepts of building design”, which was carried out as the interdisciplinary workshop entitled Punkt Podlehnik in the 2017/18 academic year, incorporated several experimental teaching methods and new principles. Integrating the highest level of interdisciplinarity according to Table 1, this course showed the compliance with the circular education scheme at UM FGPA (Figure 2, Figure 14).

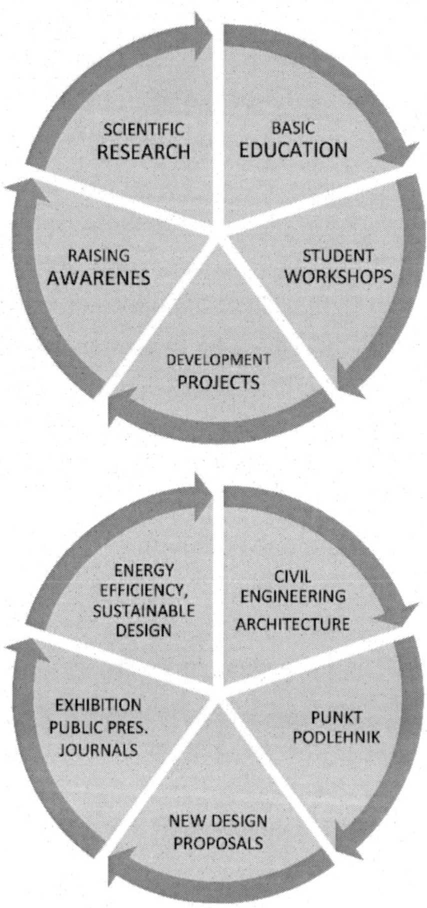

Source: Archive UM FGPA.

Figure 14. The circular education scheme at UM FGPA (left) and the education concept of the presented case study (right).

The case study proved that the educational process implemented within the course correlates to activities of the circular education scheme presented in Figure 14. The elementary knowledge was based on architecture and civil engineering, and continuously upgraded by other disciplines within the interdisciplinary workshop process. The scientific findings incorporated into the theoretical and practical part were predominately related to energy efficiency and sustainable design. The new design proposals developed by the students fit into the section of new development projects. And finally, the awareness on the topic of sustainable built environment was raised through public presentations, exhibitions, and publication in different journals and social media.

The educational process of the course integrated various disciplines related to the design of sustainable built environment. The knowledge originating from fundamental disciplines and scientific research may be distinguished from the knowledge and experiences gained from the non-academic environment (Table 2).

Table 2. Integrated disciplinary knowledge

Integrated disciplines/Source of knowledge & experiences	Basic knowledge	Scientific findings	Non-academic environment
Urban planning	+	+	
Architecture	+		
Landscape design	+		
Sustainable building design	+	+	+
Transportation engineering	+		
Civil engineering	+		
Timber construction	+	+	
Energy efficiency	+	+	
Economy viewpoint			+
Human-centred design			+

The single disciplines or expert areas shown in Table 2 were interrelated through the whole process. It is interesting that some topics were not introduced in the academic curricula at all; therefore, the collaboration with the local environment was highly productive in the sense of the comprehensive level of new knowledge and experience.

A more accurate analysis of the presented case study revealed several advantages in comparison to traditional education courses:

- students were actively involved throughout the project;

- the holistic framework of sustainability approaches was integrated in the course;
- the theoretical background was directly implemented in the project work;
- students acknowledged that they preferred teamwork to individual work, since they found it more motivating and collaborative;
- interaction between students of different disciplines, students and teachers from different expert areas, and between the academic and non-academic environment opened new challenges and perspectives of the learning process.

However, the project-based learning in the workshop exposed certain weaknesses:

- the amount of work input was not equally distributed among the team members, which was hard to control by supervisors;
- the final grades were based on the project work quality and the students' theoretical knowledge, whereby the project work was rated equally for all the team members due to the above stated remark, while in the theory knowledge segment, the students showed greater divergences. Consequently, the students claimed that the final grades were not always in line with the real contribution individual team members devoted to the project;
- a lot of organisational tasks, such as organising joint terms for different supervisors and stakeholders, site visits, communicating to municipality representatives, etc., had to be carried out by course leaders;
- demanding time management.

It would be sensible to continue experimenting with interdisciplinary problem-based learning, whereby the evaluation of student work should be improved. Excluding the above stated weaknesses, the teaching principles which employ a holistic framework shown in the presented case study reflect new approaches to education on sustainable built environment, which empower students to gain in addition to the fundamental theoretical knowledge, various experiences related to interdisciplinary collaboration, bridging the gap between traditional education and real-life experience.

CONCLUSION

Emerging challenges regarding sustainable development goals require a holistic comprehension of the global situation in which education on high-quality built environment is an important part of the mosaic. To reach comprehensiveness in knowledge, it is necessary to consider various key perspectives, which is possible through the implementation of interdisciplinary collaboration already at the educational stage. This chapter describes the experience on the interdisciplinary course "Sustainable concepts of building design" included in the education programmes of architecture and civil engineering held at the University of Maribor, Faculty of Civil Engineering, Transportation Engineering and Architecture. The aim of the course was to educate young academics and future professionals on sustainability approaches to the design of the built environment that is subject to rapid environmental, social and economic changes. The course was taught by professors of various disciplines and carried out as an interdisciplinary workshop partly implemented in the academic and partly in the local environment. The interdisciplinarity was primarily achieved with collaboration between teachers and students of different disciplines, and also through collaboration to stakeholders from the non-academic environment. A number of teaching principles were integrated in, and experimented with, in this course to achieve high effectiveness in education. The course revealed several advantages in comparison to traditional mono-disciplinary education. On the other hand, it also showed that further pedagogical innovation and improvement is still recommended to obtain the optimal results. The findings of this chapter are expected to contribute to the development of similar education approaches in academic institutions, and to the general improvement in the quality of education on sustainable built environment.

REFERENCES

Annan-Diab, F., Molinari, M. (2017). Interdisciplinarity: Practical approach to advancing education for sustainability and for the Sustainable Development Goals, *The International Journal of Management Education,* 15 (2, B): 73-83. doi:10.1016/jme.2017.0 3.006.

Becerik-Gerber, B., Gerber, D. J., Ku, K. (2011). The pace of technological innovation in architecture, engineering, and construction education: integrating recent trends into the curricula, *ITcon* (16): 411-432. http://wwwcong/2011/24.

Bozkut, E. (2016). Integration of theory courses and design studio in architectural education using sustainable development, *ERPA International Congress on Education 2015, SHS Web of Conferences,* 26 (01102). doi:10.1051/shsconf/20162601102.

Christie, B.A., Miller, K.K., Cooke, R., White, J.G. (2013). Environmental sustainability in higher education: how do academics teach? *Environmental Education Research*, 19(3): 385-414. doi: 10.1080/13504622.2012.698598

Cincera, J., Biberhofer, P., Binka, B., Boman, J., Mindt, L., Rieckmann, M. (2018). Designing a sustainability-driven entrepreneurship curriculum as a social learning process: A case study from an international knowledge alliance project, *Journal of Cleaner Production,* 172: 4357-4366. doi:10.1016/jlepro.2017.05.051.

De Gaulmyn C., Dupre K. (2019). Teaching sustainable design in architecture education: Critical review of Easy Approach for Sustainable and Environmental Design (EASED), Frontiers of Architectural Research, 8(2) 238-260. doi:10.1016/jar.2019.03.001.

Douvlou E., (2006). Effective Teaching and Learning: Integrating Problem based Learning in the Teaching of Sustainable Design, *CEBE Transactions*, 3(2): 23-37 (15). doi:10.11120/tran.2006.03020023.

Eagan, P., Cook, T., Joeres, E. (2002). Teaching the importance of culture and interdisciplinary education for sustainable development, *International Journal of Sustainability in Higher Education*, 3(1): 48-66. doi:10.1108/14676370210414173.

Li, N., Chan, D., Mao, Q., Hsu, K., Fu, Z. (2018). Urban sustainability education: Challenges and pedagogical experiments, *Habitat International,* 71: 70-80. doi:10.1016/jbitatint.2017.11.012.

Segalàs, J., Ferrer-Balas, D., Mulder, K.F. (2010). What do engineering students learn in sustainability courses? The effect of the pedagogical approach, *Journal of Cleaner Production*, 18(3): 275-284. doi: 10.1016/jlepro.2009.09.012.

Tejedor G., Segalàs J., Rosas-Casals, M. (2018). Transdisciplinarity in higher education for sustainability: How discourses are approached in engineering education, *Journal of Cleaner Production,* 175: 29-37. doi:10.1016/jlepro.2017.11.085.

United Nations. (2015). *Transforming our world: the 2030 Agenda for Sustainable Development Resolution,* General Assembly.

Vila, L., Perez, P., Morillas, F. (2012), Higher education and the development of competencies for innovation in the workplace, *Management Decision,* 50(9): 1634-1648. doi:10.1108/0025174 1211266723.

In: An Interdisciplinary Approach … ISBN: 978-1-53617-302-4
Editors: V. S. Klemenčič et al.

Chapter 2

CREATING A LIVEABLE NEIGHBOURHOOD

Vanja Skalicky Klemenčič*, PhD
University of Maribor, Faculty of Civil Engineering,
Transportation Engineering and Architecture, Maribor, Slovenia

ABSTRACT

This paper introduces a comprehensive set of urban design criteria for creating liveable residential neighbourhoods as a toll for urbanist, architect and policy makers. The design of liveable residential neighbourhoods requires a multidisciplinary approach ranging from urbanism, architecture, urban design, landscape architecture, environmental psychology to transportation and civil engineering. The main objective of the research is to define a system of criteria with a holistic approach for high-quality residential environments – not just building design but the design of residential environment as a whole. Creating liveable neighbourhoods is a complex concept; therefore, the criteria needs to be discussed from different aspects. Designing high-quality residential neighbourhoods ensuring social well-being and a healthy living environment. The established criteria are based on the Scandinavian experience, while Scandinavian countries are historically characterised by a highly developed housing culture and have been an example to many other countries. The

* Corresponding Author's Email: vanja.skalicky@um.si.

criteria are discussed in detail and implicated within the pedagogical process as part of the students' project. The paper emphasises the role of high-quality open space and green areas, which significantly contributes to liveability – human-centred urban design.

Keywords: neighbourhood, liveability, urban design, criteria, open space and green areas

INTRODUCTION

The 21st century has been characterised by extensive urban population growth and urban extension with global implications for future sustainable development. Addressing new and increasing demands for the environment, infrastructure and energy resources require a comprehensive understanding and useful approaches to the city as a link between the environment, built infrastructure and human communities through interdisciplinary cooperation (Li et al. 2018). Sustainable development is frequently associated with technological concepts (energy consumption, pollution, waste, and others); however, these are mere physical components of sustainability. So-called green technology alone does not create a sustainable society. Sustainable urban development must be socially and economically stable, while preserving a good ecological profile (Butters 2004).

The planning of a living environment does not only encompass the planning of housing, but also architectural and urban design of the narrow and wide living environment (Stanovanja, kvaliteta bivanja in razvoj poselitve - Prostor SI 2020 2000). Residential neighbourhoods have been identified at two complementary levels, i.e., as sociological and physical structures: building design, the distribution of buildings and the environmental interactions of buildings with the physical environment, facilitating user activities based on character, social qualities, and social and economic interactions in the environment with the direct community and general society (French 2012). A comprehensive approach to residential neighbourhoods is a prerequisite to creating high-quality residential neighbourhoods that support high-quality life. Highly developed housing culture with its comprehensive planning approach is typical of Scandinavian countries. Scandinavian countries have been characterised by such housing culture and a strong bond between people and nature throughout history.

Towards the end of the 1930s, i.e., following the onset of the global crisis that also affected Scandinavian countries, the latter were characterised by a new optimistic approach. They attempted to raise the general housing standard. Influenced by modernism, Scandinavian countries developed methods to attain this objective, which did not only concern housing design, but also urban design of residential environments (Nordic Associations of Architects 1978). After World War II, Sweden became an interesting target of studies for architecture of all nationalities, since the war did not affect it and the country continued to develop after the war (Pemer 2001). Slovenian architects were mainly influenced by Swedish approaches to housing development with an emphasis on Swedish models which had radically changed the views of the housing culture and the related attitude to open public space (Skalicky and Sitar 2012). Slovenian architects L. Humek and F. Ivanšek emphasised certain quality criteria in urban planning, which are still valid today, and exposed interdisciplinarity; co-operation between the architect as a generator of ideas, and sociologists, engineers, doctors, pedagogues, traffic experts etc., the diversity of building typologies, the successful integration of buildings with nature, the design of public space and green areas as the focal point of neighbourhood, and concern for a human being as an individual and a member of the community (Humek 1952; Ivanšek 1955).

Scandinavian practice of sustainable urban development has been developing and upgrading positive experiences in the planning of high-quality residential environments for decades. According to the OECD[1], a high living standard is typical of all Scandinavian countries. Regarding housing, Norway ranks first and Sweden comes in fifth in Europe (Oecdbetterlifeindex.org. 2019). The housing standard in Norway is higher than in most European countries (Hansen, 2004). In Scandinavian countries, contemplating an innovative and sustainable built environment and how to

---

[1] The Organisation for Economic Co-operation and Development (OECD) is an international organisation that works to build better policies for better lives.

develop liveable, smart and sustainable cities is crucial. The overall goal is to create added value through Nordic cooperation with the vision that Nordic countries should be the world's leading region in the area of sustainable growth and innovation (Finnsson 2015).

International practice applies several rating systems for evaluating the environmental impact of buildings on the one hand and various urban indicators for sustainable cities on the other. A few approaches to evaluating residential neighbourhoods are available; they are either a framework for planning principles in general or support the measuring of environmental sustainability through criteria (LEED-ND). Not much research at all has been done that examines the role of liveability and social sustainability in LEED-ND neighbourhoods (Szibbo 2016). A liveable neighbourhood may be defined as one that is "pleasant, safe, affordable, and supportive of human community" (Wheller 2001). Public open spaces – parks and green areas are important built environment elements within neighbourhoods, which prompt a wide range of physical activity behaviours (Koohsari et al. 2015).

In contrast to the above-mentioned tools, this research focuses on the detail explanation of a system of criteria to create liveable residential neighbourhoods in a comprehensive manner. Residential neighbourhoods are discussed from many different aspects and as a whole based on the Scandinavian experience and the original deductive method. The author proposed a hypothesis: Creating liveable residential neighbourhoods does not only concern building design, but also high-quality open space and green area design. The purpose of the research is to point out a high share of criteria that refer to open space and green area design, which in turn significantly affects the quality of life or social well-being.

THEORETICAL BACKGROUND

Essential orientations and values for the development of quality residential environments are the principles of new urbanism and smart growth transit-oriented development; a lively, safe, inviting, and healthy environment for pedestrians and cyclists, accessible to all social groups, and focused on strengthening social functions in urban spaces, strengthening identity, etc. (Urban sustainability issues 2015). A residential environment includes apartments, residential buildings, public facilities and open space where residents move, meet their needs, and carry out various activities on a daily basis; a residential environment must be defined by spatial and social indicators (Stanovanja, kvaliteta bivanja in razvoj poselitve - Prostor SI 2020 2000).

In the mid-1920s, an American urban planner C. Perry, formulated six principles for planning neighbourhood communities where the needs of a family life will be met to a greater extent: size: a residential unit development should provide housing for that population for which one elementary school is ordinarily required (its actual size depends on its population density); boundaries: the unit should be bounded on all sides by arterial streets sufficiently wide to facilitate its bypassing, instead of penetration, by through traffic; open spaces: a system of small parks and recreation spaces planned to meet the needs of any neighbourhood; institution sites: sites for the school and other institutions grouped around the centre point or a common public space; local shops: one or more shopping districts; internal street system: the unit should be provided with a special street system – the street as a whole facilitates circulation within the unit and discourages its use by through traffic (Mušič 1980).

Swedish designers embraced the philosophy of a neighbourhood or the division of larger cities into smaller units upon their expansion; of community centres for social activity development, shops and services; and of independent, new satellite cities, and developed it further. (Rudberg 1998). As mentioned earlier, similar ideas as the so-called "neighbourhood unit "concept were promoted by C. Perry in the United States. While American and British permutations were presented as direct successors of the garden city, neighbourhoods in Sweden are planned as a continuation and expansion of functionalism. Following reassessment and self-criticism in the late 1930s and at the beginning of the 1940s, the way how such housing is organised to create groups of various sizes and public spaces of different experiential quality. Interaction between a private home and public spaces became the main subject of experiments (Creaght 2011). The exhibition in Stockholm in 1930 is a crucial turning point in the modernisation of Sweden. Unlike other exhibitions of modernism, the exhibition in Stockholm was clearly focused on housing intended for the wider public. Another factor that made Sweden interesting was that its non-involvement in World War II facilitated uninterrupted construction which rapidly grew after 1945 (Rudberg 1998). Many Scandinavian and other architects worked in Sweden which was not affected by war either during or after World War II. With its strong economic position in the post-war period marked by mass housing, Sweden's labour market was attractive (Caldenby and Wedebrunn 2010).

One of the first examples of planning based on the concept of a residential neighbourhood as the ideal model of modern life in Sweden was Årsta (1946–1953). The goal was to integrate residential buildings with numerous social activities. Special attention was paid to open space. Park areas were designed according to the principle of the Stockholm school, i.e., to preserve unspoilt nature. The next level in the development of a suburban environment is the so-called ABC-town with the first example being Vällingby. While neighbourhoods were built for 10,000 residents, ABC-towns were planned for 20 to 25,000 people. The main idea was that a suburban spatial unit would bring together residential buildings with service, administrative and cultural programmes and jobs. A bond with nature remained the main idea of spatial planning (Figure 1) (Nilsson 2006). Vällingby attracted worldwide attention as an example of the "Swedish model" and a Swedish welfare state. It explicitly attested to the vision of a small country with high social ambitions. It was also an expression of modernity and nothing in Sweden was as modern as Vällingby. It became a synonym for successful urban planning and social housing (Rudberg 1998).

Even today, Scandinavian countries with the Scandinavian model appear at the top of international sustainable development charts. The Scandinavian model of a city supports a high standard of urban life and has multifaceted infrastructure that ensures sustainability. It is not solely the built environment that allows residents to live a fulfilling life; it is also the economic, social and cultural infrastructure (Sheriff 2015). Rambøll's "Model for assessing liveable cities" within the "Nordic Built Cities" programme (Finnsson 2015) defines the term "liveability" as creating conditions for a decent life for all residents of cities, regions, and communities, including their physical and mental health. An environment that is pleasant for living is based on the principle of sustainable and smart development, making it sensitive to nature and the protection of natural resources. In addition, Scandinavian architect and urbanist Jan Gehl is a global leader in people-centred urban design in the world (Gehl 2019).

Figure 1. The centre of Vällingby with social activities and the placement of buildings in the natural environment.

Figure 2. Contemporary liveable neighbourhoods of Pilestredet Park in Oslo (right) and Hammarby Sjöstad in Stockholm (left).

Contemporary neighbourhoods of Pilestredet Park in Oslo and Hammarby Sjöstad in Stockholm, which received several international awards and originated from the first two decades of the 21st century with mixed use follow the principles of sustainable development with an integrated approach. Both neighbourhoods were also included in the process of the development of criteria as case studies. Pilestredet Park is an example of the urban transformation of a city centre, whereas Hammarby Sjöstad is an example of new urban development on brownfield former industry sites. While designing both residential neighbourhoods, the quality design of open space and green areas is of key importance in providing the recognisability of residential environments and of liveability – physical and mental well-being of the dwellers (Figure 2).

METHODS

An innovative deductive methodology serves as a platform for developing a set of criteria for creating liveable residential neighbourhoods, and ensuring healthy, pleasant, and attractive environments for people. The criteria are based on the contemporary Scandinavian planning principles. The overall framework is established on various aspects of liveability, including various key objectives and values, sets of criteria, and indicators. The methodology is based on various phases which consist of: the establishment of crucial aspects, the development of objectives and values, the selection of spatial solutions and measure data indicators (based on Scandinavian case studies Pilestredet Park and Hammarby Sjöstad), the development/identification of liveability criteria and a detailed definition of each criterion and the verification of the method (using case studies from Slovenia). These phases are presented in more detail in the article entitled "Comprehensive assessment methodology for liveable residential environments" (Skalicky and Čerpes 2019).

This article presents a set of criteria for creating liveable neighbourhoods, and a detailed definition of each criterion as well as their use in the pedagogical process within the students' concept of a residential neighbourhood design as is shown in Figure 3. A detailed definition of criteria is based on research conducted by various experts in different disciplines; from urbanism to environmental psychology, who studied urban design and interactions between a physical form of a residential environment and its users.

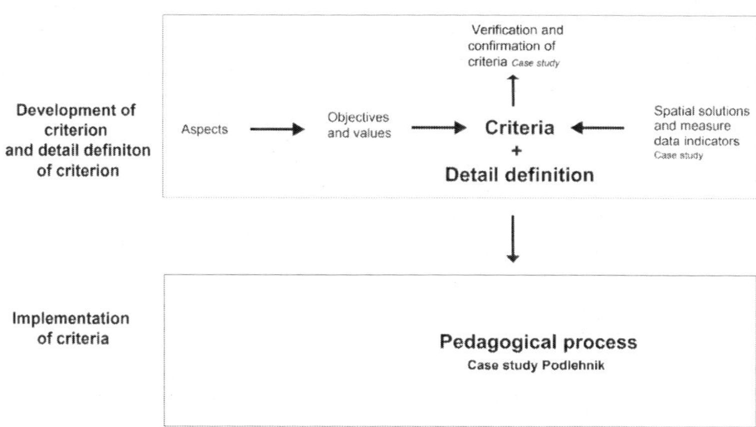

Figure 3. Method model.

Figure 4. Objectives and values for designing liveable neighbourhoods. Adapted from (Skalicky and Čerpes, 2019).

Table 1. Set of liveability criteria

Objectives and values	Criteria
	Environmental aspect
Compact and effective growth	Prevention of urban sprawl and impact on surrounding natural environment
	Regeneration within the existing city boundary and development of brownfields
	Closer proximity to the various amenities of everyday life
	Promoting walking, cycling, and use of public transport
Protection of natural resources	Reduction of energy
	Supply of energy from renewable sources
	Improving efficiency in energy supply and distribution by installing district energy networks
	Decreasing light pollution and energy consumption, necessary for lighting
	Local use of rainwater and storm water
	Protection and enhancement of water resources, green areas, vegetation
Waste and recycle management	Recycled and on-site reused building material
	Waste recycling and use of combustible waste for fuel in energy cogeneration
	Sociological aspect
Sense of belonging	Distinction between private and public space; hierarchy of open spaces in a residential environment
	A residential environment for leisure time and stronger social interactions
	Well-maintained residential environment
	Involvement of residents in the residential environment
	Awareness, participation, and education of residents – active role of dwellers in the residential environment

Objectives and values	Criteria
Sense of safety	Lively and controlled public space
	Protection of private space and interaction with public space
	Design of residential environment, enabling orientation in space
	Lighting of open space in the residential environment
	Priority for pedestrian and traffic safety
Sense of enjoyment and comfort	Human scale of the residential environment
	Reducing noise level and creating quiet areas
	Reducing the heat island effect
	Functional aspect
Accessibility	Integration of different social and age groups into the residential environment
	Universal design: designing open public spaces accessible to all
	Open space, i.e., social space
Integration into a wider urban structure	Residential environment connecting dwellers and other city residents
	Promoting movement: walking and cycling
Flexibility of residential environment	Temporary use and shared use of space
	Cultural aspect
Context and identity	Engagement with urban context
	Protection of natural and cultural heritage
Human-oriented environment	Recognisability of residential environment
	Walking-friendly and lively residential environment
Attractiveness and readability	Empirical and sensory rich residential environment

Adapted from (Skalicky and Čerpes, 2019).

The basis for the formation of criteria is a set of objectives and values that stem from a Scandinavian source and literature analysis and represent the framework of a system of criteria classified into different aspects for an integrated approach to liveable residential neighbourhoods. The discussion of a residential environment builds on twelve objectives and values that define various aspects for creating a living environment with first-class ambiance that is attractive for people and sensitive to nature and space is shown in Figure 4. The environmental, functional, sociological, and cultural aspects were included in the research model.

The 36 liveability criteria refer to the scale of a residential neighbourhood. As a residential neighbourhood is integrated into a wider urban structure, it requires the verification and comprehension of the wider urban structure. The criteria refer to different elements of a residential neighbourhood: buildings, transportation, supply, programme, and open space and green areas. The paper presents a systematic review of the liveability criteria as shown is Table 1, followed by a detailed description of each criterion.

LIVEABILITY CRITERIA

Compact and Effective Growth

Prevention of Urban Sprawl and Impact on Surrounding Natural Environment

As the urban population grows and efforts for high-quality life intensify, ensuring sustainable use and treatment of space is an important aspect of planning. The process of combining residents and their activities reduces the use of space for settlement, preserves and promotes biodiversity, and prevents natural resources from depleting.

Regeneration within the Existing City Boundary and Development of Brownfields

The development of new residential environments is based on densification within built-up areas, the vicinity of public transport, and developments in degraded areas. Compression should not interfere with existing green areas on the one hand and enable each resident to live in their immediate vicinity, i.e., five to ten-minute walk, on the other (Nelson 2005).

Closer Proximity to the Various Amenities of Everyday Life

Residential environments should be planned to facilitate better access to various goods and services: crucial public functions and services, job opportunities, spending leisure time, and transport, thus providing a higher quality of life.

Promoting Walking, Cycling, and Use of Public Transport

Sustainable mobility, which contributes to reducing harmful emissions, noise levels, energy source consumption, and investments in expensive infrastructure, includes walking, cycling and using public transport as well as reducing the scope and use of road traffic. Urban planning must comprise sustainable transport to enable people to cross a city simply and cheaply with a low environmental impact, and at the same time, to facilitate the movement of people, goods and services.

Protection of Natural Resources

Reduction of Energy

The general objective is to reduce the consumption of thermal energy and electricity. The energy system is divided into two sets: energy demand, i.e., the sum of the whole energy load, and energy supply, i.e., how the energy is supplied. Energy demand is lower in energy efficient buildings.

Reduced energy demand facilitates easier supply of a high share of renewable energy obtained in the vicinity (Fraker 2013). To reduce energy demand, we need to strive for passive building systems (Kvalitetsprogram Filipstad områderegulering 2012). The main role in the optimisation of energy efficiency is held by the orientation of a building, which is related to the design of the area and building morphology (De Matteis 2010). Rational behaviour of dwellers is also crucial to reduce power consumption.

Supply of Energy from Renewable Sources

Energy supply in residential environments primarily envisages the use of local, innovative, low-carbon, renewable energy sources. When dealing with energy supply, Scandinavian cities are particularly innovative. They use local sources and solutions: wind, waste-to-energy, biogas from sewage sludge, and biomass to reduce the consumption of renewable energy sources and carbon emissions, and impact on climate change with decarbonisation.

Improving Efficiency in Energy Supply and Distribution by Installing District Energy Networks

Traditional and centralised energy production is frequently inefficient, as 60 per cent of energy or more is lost. Power generation from fossil fuels or nuclear energy may release large quantities of thermal energy which is discarded. In the past 100 years, Scandinavian cities have provided solutions with decentralised, district energy networks, systems that produce large quantities of energy but are closer to power consumption sites, which foster the use of waste heat and energy (Nordic Solutions for Sustainable Cities 2012).

Decreasing Light Pollution and Energy Consumption, Necessary for Lighting

Urban spaces are exposed to light pollution with excessive energy consumption, which affects health, well-being and nocturnal biodiversity.

Local Use of Rainwater and Storm Water

Better rainwater and stormwater management prevents urban drainage systems from overloading and related possibility of floods. Rainwater and stormwater use and harvesting are one of the main elements of the urban landscape design in residential environments. Water is no longer an undesired element to be removed from buildings and open spaces as soon as possible, but it has become part of designed outdoor areas (Pilestredet Park 2009).

Protection and Enhancement of Water Resources, Green Areas, Vegetation

High-quality water and green areas enable residents to use them for recreational purposes, and preserve the habitat of various organisms. Biodiversity and ecosystems must be preserved.

Waste and Recycle Management

Recycled and on-Site Reused Building Material

Sustainable building strives for lower consumption of harmful materials and sources during construction and building operation, while promoting the use of recycled building materials. In this way, it reduces the quantity of waste material and contributes to a cleaner environment. Sustainable building also includes planned demolition and disintegration at the end of the life cycle in order for waste materials to be separated, and suitable and harmless materials to be recycled and reused.

Waste Recycling and the Use of Combustible Waste for Fuel in Energy Cogeneration

Energy generation from renewable sources, particularly from organic waste (food, sewage, green waste) constitutes a complete circle, whereby one system's waste is another system's energy source (Fraker 2013).

Sense of Belonging

Distinction between Private and Public Space; Hierarchy of Open Spaces in a Residential Environment

Newman (1996) defines a strong connection between a clearly defined territorial belonging and safety. On the basis of research, he explains that dwellers maintain and supervise only spaces that are defined as theirs. Other spaces are unsupervised and without belongings. Newman further maintains that dwellers identify themselves with space and have a sense of belonging to the residential environment only when it is linked with the residential building. Clearly defined spaces and related belonging are crucial prerequisites for social opportunities, and for interaction with others on the one hand and the protection of private space on the other. Planning spaces in between, semi-private and semi-public, fosters a gradual and soft transition between public and private spaces, increases the likelihood of opportunities for contacts between areas, and provides dwellers with the option to choose a contact and protect their privacy. Safety and the ability to understand space enhance clear physical demarcations of spaces in a residential environment, and between a residential environment and other parts of a city. Dwellers thus perceive a residential environment as their own, while visitors are aware that they are mere guests in this space (Gehl 2010). A clear definition of different types of space in a residential environment affects a sense of safety and consequently, dwellers' sense of belonging.

A Residential Environment for Leisure Time and Stronger Social Interactions

Open public spaces are spaces where people may spend their leisure time, recreate, gather, and socialise. Accessibility, contents and use-related equality of people are vital. Their arrangement in space, orderliness and programme diversity are important. They enable people to spend their active and passive leisure time in relation to their home, and meet their social needs, they bring residents together and significantly affect their attitude to the residential environment. The connection of dwellers with the residential environment and a sense of community among them largely depend on the planning and use of public space. The latter must foster various experiences and activities, and the well-being of various age groups. The most important for children are playgrounds and their equipment, while for young people, the most important are sports grounds where they may socialise away from people who are older and younger than them; a space is pleasant for older people when they may walk, sit, relax, and talk in peace (Polič 2007).

Well-Maintained Residential Environment

Well-maintained public space is an important part of a residential environment, which affects its reputation and contributes to the economic value of real estate. Such an environment gives dwellers a stronger sense of belonging and makes it easier for them to identify themselves with it. Only orderly and suitably maintained spaces condition a sense of belonging and attractiveness of public open spaces for use (Van Kemmpen et al. 2007). It is interesting to compare multifamily residential buildings where spaces in the building, which are only shared by few apartments (several entrances), are well-maintained and orderly, and common spaces which may be shared by many families in the building. The same applies to the image of outdoor public spaces (Newman 1996).

Involvement of Residents in the Residential Environment

According to Trstenjak (1984), the main function of a home is to provide people with a sense of safety and privacy. People who live alone need particularly to be included in activities and in the "attitude to the environment." By including dwellers in various activities, lonely persons make the necessary contact, have conversations, meet people, receive news, and experience the necessary change of the environment so that they do not become resigned to monotony, boredom and isolation. In this way, they maintain the awareness that society still needs them. Trstenjak illustratively demonstrates isolation on the example of large shops which are not pleasing in terms of ecopsychology; people desire quite the contrary: they desire personal not merely mechanical service, and wish to consult and talk.

Awareness, Participation, and Education of Residents – Active Role of Dwellers in the Residential Environment

Participation and information of dwellers are important conditions for the successful planning, maintaining, and refurbishing of a residential environment, while education contributes to its comprehensive and sustainable use. The participation of dwellers in managing and maintaining a residential environment increases the likelihood of a successful activity and a sense of belonging.

Sense of Safety

Lively and Controlled Public Space

Streets for pedestrians are the most important spaces of a living environment and its most vital part intended not only for passive but also active use. While a frequently used street is probably safe, a deserted street is likely less safe or dangerous. To achieve a sense of safety, users must be present in streets, they observe the space and stir interest of a sufficient number of people in houses along the street, who observe activities in the street too. The frequency of people in a residential environment fosters social control, providing a sense of safety (Jacobs 1961). Newman (1996) states that people in residential buildings in which a common hallway is only shared by few apartments know each other and recognise strangers or intruders more easily, which significantly contributes to a sense of safety and belonging. Insecure residential environments which do not provide safety are lifeless streets; monofunctional buildings devoid of activities during the day; closed, lifeless and dark façades; unsuitably lighted open spaces, deserted paths, pedestrian tunnels, dark nooks and crannies, and too many bushes (Gehl 2010).

Protection of Private Space and Interaction with Public Space

Newman (1996) illustrates how different forms of building typologies constitute various types of outdoor open space, and facilitate diverse organisation of access to buildings and integration in the residential environment.

Design of Residential Environment, Enabling Orientation in Space

It is indicative of good urban design when users easily and without problems finds the way to their destination. The organisation of space should be such that each link in a street network has a clear visual characteristic, space has a distinctive character, and that important streets can be distinguished from less important ones, enhancing a sense of safety while walking through a residential environment (Gehl 2010).

Lighting of Open Space in the Residential Environment

Open space lighting is crucial to designing a safe environment, contributing to the quality of life.

Priority for Pedestrian and Traffic Safety

In the last 50 years, the planning of cities has conformed to road traffic with larger areas for such traffic and parking. This has made walking more difficult and less attractive, and the number of road accidents has grown. Transportation planning has changed significantly in the 21st century, particularly in Europe. Walking and cycling take priority, and better understanding of the causes of road accidents fosters the development of new approaches to transportation planning. In the 1960s, the first pedestrian street was introduced in Europe. Only two street models were known, i.e., for road traffic or pedestrians. Many different types of streets have been developed so far, which provides planners with diverse opportunities: roads only for road traffic, roads in traffic-calming zones (30 km/h), roads with pedestrian priority, traffic-calming zones (15 km/h) etc. (Gehl 2010).

Sense of Enjoyment and Comfort

Human Scale of the Residential Environment

Scale is a crucial factor to how people experience the environment around them. Human scale refers to dimensions and shapes, and other physical elements that correlate with human scale (Grenaker 2105). Dimensions of residential buildings, i.e., height and length, directly affect the size of outdoor space between buildings, since good lighting and exposure to sunlight require a suitable distance between buildings. People do not like too large, too narrow or small and completely closed outdoor spaces in which they are lost, which do not provide them with enough intimacy, and are too shadowy etc. People need a correctly dimensioned open space that fits well to them. The lowest threshold of dimensions of an indoor (open) space, which supports good lightning and exposure to sunlight, requires a certain distance between buildings which is usually twice the height of buildings (Jernejec 1974). According to Gehl (2010), an important element when planning urban spaces is a link between the senses, communication, and the dimensions of the space, i.e., the social field of vision.

Reducing Noise Level and Creating Quiet Areas

Reducing the speed limit to reduce of noise and vibration levels. Placing housing and certain public programmes at sufficient distance from noise and pollution sources.

Reducing the Heat Island Effect

Microclimate determines the way open spaces are used. It particularly affects the typology of open space frequented by pedestrians and cyclists who need a comfortable and healthy environment. In view of the predicted climate change as a consequence of global warming, more frequent heat waves and heat islands in residential environments may be expected. Lawns reduce air temperature in their immediate vicinity in the summer. Green areas tend to warm up less than hard surfaces (asphalt, concrete, macadam etc.), facilitating air circulation: cooler air from green areas pushes away warm air from urban areas. Trees give shade, cooling lower buildings whose southern and western sides are most exposed to solar heat (Jernejec 1974).

Accessibility

Integration of Different Social and Age Groups into the Residential Environment

For a residential environment to correspond to various types of families and other user groups, it must encompass diverse housing types of various prices, shapes, and sizes (Kvalitetsprogram Filipstad områderegulering 2012). Residential buildings must correspond to users who use them over various periods and within various time frames: long-term use by families, students use them during the year, long-term or seasonal use by hostels; residential buildings must correspond to the size and type of households (housing for single persons, families with/without children); housing for persons with special needs (disabled, older people).

Universal Design: Designing Open Public Spaces Accessible to All

Open public space and pedestrian zones facilitate safe and smooth access and accessibility for everyone regardless of age and abilities.

Open Space, i.e., Social Space

Each residential area is inhabited by various groups of people with corresponding diverse needs. Open space, such as so-called social space, provides dwellers and other residents with different contents intended for recreation and the spending of leisure time (Jernejec 1974).

Integration into a Wider Urban Structure

Residential Environment Connecting Dwellers and Other City Residents

A residential environment must be designed to support the access to, and use of, open public space and other public contents not only by dwellers but also other residents of a city, visitors, and tourists.

Promoting Movement: Walking and Cycling

Planning is based on the preservation of the natural environment from the aspect of environmental protection and human health. Promoting walking and cycling fosters a healthy lifestyle and well-being, reducing health care costs. One of the essential tasks is to link green areas and recreational opportunities for residents of a city, while promoting biodiversity. There are certain documents which point out the significance of green areas, particularly in relation to the construction and expansion of cities, e.g.: "Stockholm Park program" (Nelson 2006) explicitly emphasises the right of residents to live in the immediate vicinity of green areas, and defines radius for access to green areas. Physical inactivity results in decreased quality of life, extreme increase in health care costs, and a shorter lifespan (Gehl 2010).

Flexibility of Residential Environment

Temporary Use and Shared Use of Space

As the project unfolds, the area should provide the wider population with availability and recognisability with temporary use and activities in buildings and outdoor areas (Kvalitetsprogram Filipstad områderegulering 2012).

Context and Identity

Engagement with Urban Context

Local context is used to mark specific building typologies, solutions and the use of materials on the one hand, and the establishment of a sensible relationship between buildings and the area on the other (Leupen and Mooij 2011). Urban design is based on the existing characteristics of an area included in the process of new design. Including context in the preparation of new proposals means recognising links to existing urban tissue, and creating functional and design-related links to existing space.

The continuity of existing urban tissue is reflected at the level of urbanism and architecture. When building their homes, people traditionally used local materials that were available, widely used, and known. The construction was based on local tradition and habits.

Protection of Natural and Cultural Heritage

Elements of natural and cultural heritage of a residential environment build the identity of an area.

Human-Oriented Environment

Recognisability of Residential Environment

The way apartments in a residential building are organised, the size and form of the building, and the way they are combined into groups create unique characteristics. In a way, the element that links a group of buildings is a well-maintained outdoor space. Residential buildings are interconnected in diverse and numerous ways (Jernejec 1974). The recognisability and identity of a residential environment are created by a multitude of open spaces, various characters, dimensions, different approaches to design, and various contents.

Walking-Friendly and Lively Residential Environment

Dwellers need a safe and attractive environment to foster their socialisation and play, where road traffic is limited.

An environment will be more attractive if many people of various age groups and with different skills are present and use the space in different ways either actively or passively. A residential environment needs to be created, which will promote walking. The ground floor of a building and the space next to it play the essential role (Gehl 2010).

Attractiveness and Readability

Empirical and Sensory Rich Residential Environment

A well-designed open space manages to address a person and stimulate a comprehensive perception of the environment. Such a space fosters a rich experience which is a result of a mindful and perceptive detection and treatment of the planner. In such a space, even a person with little appreciation for designed environment notices quality and perceives it as a value (Lee 2009). The elements of a green area create outdoor spaces of various sizes and interconnect them (Jernejec 1974).

Implementation of Criteria in the Pedagogical Process: Podlehnik Case Study

Workshops for students are a form of project-based learning and cooperation between local communities and educational institutions. The location of the students' concept of a new residential neighbourhood is in the urbanised rural area of Podlehnik.

Podlehnik has been recording a decline in the number of residents since 1991. Depopulation areas encompass part of Haloze and Slovenske gorice, particularly small, the most dispersed, and poorly accessible settlements. Most of these are border settlements including the Municipality of Podlehnik. Between 2002 and 2008, certain settlements in Haloze and Slovenske gorice saw population growth, particularly local supply centres and new municipal centres, including Podlehnik[2] (Lampič and Rebernik 2011). The trend of the depopulation of rural areas is also characteristic of the Nordic region. Smaller rural municipalities struggle with depopulation, while larger municipalities are challenged by gentrification and high population density, i.e., crowding and a growing need for housing (Finnsson 2015). Life in rural areas may be promoted by planning new contemporary

---

[2] And between 2010 – 2011, in 2012 and between 2015-2016 as of July 13, 2019, the Statistical Office of the Republic of Slovenia listed on its website.

and high-quality way of living with the prerequisite being good transport connections with urban centres. The use of criteria for creating a liveable neighbourhood based on urban environments may also be applied to rural areas regarding which one of the main objectives requires non-dispersed development of settlements and high-quality residential environments, as our standpoint is that most residents of rural areas have similar needs to residents of urban areas. Unlike urban centres, rural areas are marked by lower population density, slower population growth, and space where rural identity is more distinct than urban identity (Drozg and Pelc 2008).

The students' project is a concept stage of the design of a new residential neighbourhood. The objective of the project was a comprehensive approach to the residential neighbourhood design. Initially, students were presented with detailed criteria for designing a liveable neighbourhood. The presented criteria were included by students in their concepts through their weekly consultations with mentors. Taking the criteria into account, it was noticed that over half of the criteria refer to open space and green area design, which are presented in short on the example of the students' project.

All students' concepts of a residential neighbourhood take into account the principles of compact growth and the prevention of rural sprawl, as they facilitate access to crucial public functions and green areas in the immediate vicinity. The preservation of green areas in the area in question contribute to more efficient rainwater management, prevent heat islands from forming, and support recreational and relaxation activities of users. The planning of various types of open spaces and green areas fosters social opportunities, diverse open spaces facilitate socialising and various experiences and activities for all age groups (walking, sitting, lying on the grass, picnics, play for children). In residential buildings, a common space is only shared by few dwellers, which enables closer connection and better supervision and maintenance of the space. Ground floors in multifamily residential buildings are intended for public use and create a vibrant street. The street scheme is designed clearly and transparently. The concept includes various types of streets: for road traffic, and large areas are dedicated to pedestrians. Access to the residential environment and elements of urban design clearly define entrances to the residential environment.

Building typologies summarise the immediate environment. Local materials were used. Accumulating building volumes, elements of urban design and greenery divide space into recognisable districts. Designing open space includes a cohabitation of vulnerable groups and intensifies the experience of the natural environment (e.g., the placement of trees in the atrium of a community centre interprets a change of different seasons).

CONCLUSION

The development of settlements and residential neighbourhood is focused on the protection of the environment and human health to provide residents with a higher quality of life. People need to be perceived as individuals and part of the society. The environment in which they live must nurture their mental and physical well-being as that of an individual, and foster socialisation and contact with others on the one hand and intimacy on the other. The notion of a high-quality residential environment comprises the cohabitation of various social groups whom it enables to lead a high-quality life, and prevents the isolation of the residential environment and its dwellers. Special attention must be paid to vulnerable social groups, and provide them with equal integration into, and access to, a wider social environment. Space must be designed to be attractive to users and harmonised with wider space. To strengthen the belonging to the environment and enable users to identify with it, space must have its own identity and be recognisable and interesting.

It has been established on the basis of research that a liveable neighbourhood is characterised by compact and efficient growth, the protection of natural resources, waste and waste material management, and that a high-quality residential environment fosters users' sense of belonging, safety and comfort, is available to various users, integrated in wider space, flexible, and designed in its own context with its own identity: it is a human, vibrant, attractive, and recognisable residential environment.

The research presents the system of criteria based on examples of contemporary practice in the planning of residential environments in Scandinavia with sustainable development values and empathy, and continued development of residential environments, which is reflected in human design. The results of the research confirm the significance of a comprehensive approach to the development of residential neighbourhoods, and the role of open space and green areas. Based on the synthesis of the system of criteria developed for creating a liveable neighbourhood and its implementations, we find that the designing of high-quality open space and green areas vitally contributes to the quality of a residential environment, since over half of the criteria refer to it.

The article presents in detail the urban design criteria for creating a liveable neighbourhood, which facilitate a systematic approach in the planning process of new residential environments and are tools for the analytical verification of the state of existing residential environments. Liveability criteria are particularly intended for experts, and promote interdisciplinary and intersectoral cooperation. They are also meant to educate the general public as part of the planning process, and to help understand the significance of various elements of a residential environment to achieve a higher quality of life.

In the research and during the pedagogical process, we established that creating a liveable neighbourhood is a complex task in the light of the multitude of criteria and their roles which are crucial to achieving the quality of life on the one hand and to protecting the natural environment on the other. An important role is held by the high-quality design of open space and green areas.

REFERENCES

Butters, C. (2004). A Holistic Method of Evaluating Sustainability. In *Building and Urban Development in Norway: a selection of current issues,* edited by Jens F. Nystad, 38-43. Husbanken.

Caldenby, C., Wedebrunn, O. (2010). *Living and Dying in the Urban Modernity*. Denmark: Docomomo.

Creagh, L. (2011). From Acceptera to Vällingby: The Discourse on Individuality and Community in Sweden (1931-1954), *Footprint 9:*5-24. doi:10.7480/footprint.5.2.737.

De Matteis, F. (2010). Housing Europe. In *Housing for Europe: Strategies for Quality in Urban Space, Excellence in Design, Performance in Building*, edited by C. Clemente, F. De Matteis, 45-70. Roma: DEI - Tipografia del Genio Civile.

Drozg, V., Pelc, S. (2008). Konceptualne smeri v slovenski geografiji podeželja, *Revija za geografijo* 3(2): 87-100. [Conceptual directions in Slovene rural geography, *Journal of Geography* 3(2): 87-100].

Finnsson, P. T. (2015). *Nordic Urban Strengths and Challenges – How do we perceive ourselves when it comes to developing sustainable, smart and liveable cities?* Holmen: Nordic Innovation.

Fraker, H. (2013). *The Hidden Potential of Sustainable Neighbourhoods*. Washington: Island Press.

French M. (2012). *Sustainable housing for sustainable cities: A policy framework for developing countries*. Nairobi: UN-Habitat.

Gehl, J. (2010. *Cities for People*. Washington: Island Press.

Gehl. (2019). *Jan Gehl at 80 - Gehl*. [online] Available at: https://gehlpeople.com/news/jan-gehl-at-80/ [Accessed 30 Aug. 2019].

Grenaker, L. (2015). *Densification with Quality of Urban Life*. Master thesis, Aalborg University.

Hansen, T. (2004). Housing Policy Challenges in a Country with High Housing Standards and a Marketgoverned Housing Supply. In *Building and Urban Development in Norway: a selection of current issues,* edited by J. F. Nystad, 60-65. Husbanken.

Humek, L. (1952). Po Švici, Švedski, Finski. *Arhitekt* 8:36-39. [Across Switzerland, Swedish, Finnish. *Architect* 8:36-39].

Ivanšek, F. (1955). Osnovne poteze skandinavske stanovanjske politike. *Arhitekt* 16:36-40. [Basic features of Scandinavian housing policy. Architect 16:36-40].

Jacobs, J. (1961). *The Death and Life of Great American Cities*. New York: Random House.

Jernejec, M. (1974). *Stanovanjsko okolje – 1. del: Človek, njegovo okolje in potrebe.* Ljubljana, Urbanistični Inštitut SRS. [*Residential environment - Part 1: Human, his environment and needs.* Ljubljana, Urban Planning Institute SRS].

Koohsari, M. J., Mavoa, S., Villanueva, K., Sugiyama, T., Badland, H., Kaczynski, A. T., Owen, N., Giles-Corti, B. (2015). Public Open Space, Physical Activity, Urban Design and Public Health: Concepts, Methods and Research Agenda, *Health & Place* 33:75-82. doi:10.1016/j.healthplace.2015.02.009.

Kvalitetsprogram Filipstad områderegulering. (2012). Oslo Havn. [*Quality program Filipstad area regulation*. 2012. Oslo Havn].

Lampič, B., Rebernik D. (2011). *Spodnje Podravje pred izzivi trajnostnega razvoja.* Ljubljana: Znanstvena založba Filozofske fakultete Univerze v Ljubljani. [*Lower Podravje region facing the challenges of sustainable development.* Ljubljana: Publisher of the Faculty of Arts, University of Ljubljana].

Lee, U. (2009). *SLA: C3 Landscape.* Seoul: C3 Publishing Co.

Leupen, B., Mooij, H. (2011). *Housing Design.* NAi Publishers: Rotterdam.

Li, N., Chan, D., Mao, Q., Hsu, K., Fu, Z. (2018). Urban sustainability education: Challenges in pedagogical experiment, *Habitat International 71:70-80.* doi:10.1016/j.habitatint.2017.11.012.

Mušič, V. B. (1980). *Urbanizem – bajke in resničnost.* Ljubljana: Cankarjeva založba. [*Urbanism - myths and reality*. Ljubljana: Cankar publishing house].

Nelson, A. (2006). *Stockholm, Sweden: City of water.*

Newman, O. (1996). *Creating defensible space.* Washington: U.S. Department of Housing and Urban Development Office of Policy Development and Research.

Nilsson, L. (2006). The Stockholm Style: a model for the building of the city in parks, 1930s-1960s. In *The European City and Green Space: London, Stockholm, Helsinki and St. Petersburg 1850 – 2000,* edited by P. Clark, 141-158. Burlington: Ashgate Publishing Limited.

Nordic Associations of Architects (DAL, NAL, SAR) for the U.I.A. (1978). *Nordic Housing - Habitación Nórdica – 1945/1980 - Denmark - Finland - Norway – Sweden.* Mexico.

Nordic Solutions for Sustainable Cities. (2012). The Nordic Eight. Copenhagen: The Nordic Council of Ministers.

Oecdbetterlifeindex.org. (2019). *OECD Better Life Index.* [online] Available at: http://www.oecdbetterlifeindex.org/topics/housing/ [Accessed 30 Aug. 2019].

Pemer, M., (2001). *Developing a sustainable compact city in Stockholm, Sweden.* Instanbul: Thematic Committet.

Pilestredet Park - a tale of sustainable urban development. (2009). Statsbygg.

Polič, M. (2007). *Okoljska psihologija.* Ljubljana: Univerza v Ljubljani, Filozofska fakulteta, oddelek za psihologijo. [*Environmental psychology.* Ljubljana: University of Ljubljana].

Rudberg, E. (1998). Building the Welfare of the Folkhemmet. In C. Caldenby, J. Lindvall, & W. Wang (Eds.), *20th-Century Architecture Sweden,* 111-141. Munich; New York: Prestel Pub.

Sheriff, S. (2015). *Urban Attraction: Harmony and Diversity in the City Landscape*. Nordic City.

Skalicky, V., Sitar, M. (2012). The Concepts on Quality of Life in the Maribor post-WW2 Housing Estates, *Architecture, Research* 1:18-25.

Skalicky, V., Čerpes, I. (2019). Comprehensive assessment methodology for liveable residential environment, *Cities* 94:44-54. doi:10.1016/j.cities.2019.05.020.

Stanovanja, kvaliteta bivanja in razvoj poselitve - Prostor SI 2020. (2000). Ljubljana: Urbanistični Inštitut Republike Slovenije. [*Housing, quality of living settlement development - Space SI and 2020.* Ljubljana: Urban Planning Institute of the Republic of Slovenia].

Statistical Office of the Republic of Slovenia. (2019). *Slovenian statistical regions and municipalities in numbers.* [online] Available at: https://www.stat.si/obcine/sl/2016/Municip/Index/122 [Accessed 13 July 2019].

Szibbo, N. A. (2016). Assessing Neighborhood Livability: Evidence from LEED® for Neighborhood Development and New Urbanist Communities, *Articulo – Journal of Urban Researc 14:1-24.* doi:10.4000/articulo.3120.

Trstenjak, A. (1984). *Ekološka psihologija.* Ljubljana: Gospodarski vestnik. [*Ecological psychology.* Ljubljana: The Economic Bulletin].

Urban sustainability issues – Resource-efficient cities: good practice. (2015). European Environmental Agency.

Van Kempen, R. (2007). *Regeneracija velikih stanovanjskih sosesk v Evropi: Priročnik za boljšo prakso.* Ljubljana: UIRP. [*Regeneration of large residential neighbourhoods in Europe*: *A guide to best practice*, Ljubljana: UIRP].

Wheeler, S. M. (2001). *Livable Communities: Creating Safe and Livable Neighborhoods, Towns, and Regions in California.* UC Berkeley: Institute of Urban and Regional Development. https://escholarship.org/uc/item/8xf2d6jg.

In: An Interdisciplinary Approach … ISBN: 978-1-53617-302-4
Editors: V. S. Klemenčič et al.

Chapter 3

SUSTAINABLE TRAFFIC AND THE LIVING ENVIRONMENT

Marko Renčelj*[*]*, PhD
University of Maribor, Faculty of Civil Engineering,
Transportation Engineering and Architecture, Maribor, Slovenia

ABSTRACT

In this chapter, we deal with sustainable transport in connection with the living environment. This is directly related to the interdisciplinary approach in the field of higher education in the area of sustainable building design. Today, we are very aware of the negative consequences of traffic, especially road traffic (congestions, environmental pollution, road accidents etc.). Sustainable mobility is an established approach with which we could reduce the negative effects of road traffic. Sustainable mobility does not imply just a restriction of the use of cars, but encourages and gives people the opportunity to choose (more often) other modes of transport – public transport, bicycles, walking etc. Our goal is to transform individual solutions in individual areas that combine seemingly completely different

[*] Corresponding Author's Email: marko.rencelj@um.si.

systems into an integrated transport system that is safe, healthy, and efficient.

This chapter thus contains content that – in the field of the living environment in connection with mobility – includes the following key steps: sustainable mobility, sustainable road safety, (road) infrastructure measures to ensure sustainable mobility. The final section presents examples of an interdisciplinary approach/search for solutions at the level of student tasks and master's theses.

Keywords: sustainable traffic, sustainable mobility, sustainable road safety, road infrastructure measures, 30 km/h zones, traffic calming, infrastructure for cyclists and pedestrians

INTRODUCTION

Nowadays, we are well aware of the negative effects/consequences of all forms and modes traffic. We should particularly point out the negative effects/consequences of road traffic which is the most common. In this chapter, we focus primarily on road motor traffic, mainly on passenger road vehicles.

All too often, we still choose motor vehicles (cars) for our mobility. The use of cars increases possibilities of congestions, road accidents, environmental pollution etc. Urban centres (cities, settlements) are packed with cars (in motion or stationery – parked vehicles), and even vulnerable road users (pedestrians, cyclists etc.) are frequently "victims" of drivers' inappropriate behaviour. A solution to this problem (or an option to improve the existing situation in this field) is the realisation of the sustainable traffic system/policy. In this regard, it must be initially pointed out that this does not imply the restriction of the use of cars, but encourages and gives people the opportunity to choose (more often) other modes of transport – public transport, bicycles, walking etc. Individual solutions in individual/partial fields that combine seemingly completely different systems must be transformed into an integrated transport system that is safe, healthy, and efficient.

Potential individual solutions are presented below, particularly solutions that have proven themselves in practice and are the most suitable for our environment according to the authors.

SUSTAINABLE TRAFFIC, SUSTAINABLE MOBILITY

We cannot imagine life today without mobility; it is a necessity and brings quality to our lives. One of the goals in this field is to provide a high level of accessibility, while reducing negative environmental impacts. Road motor traffic is a source of many environmental problems. In 2006, the European Commission approved the document "*Thematic strategy on the urban environment*" (EU, 2016), which points out sustainable mobility plans for cities as a response to the problem of urban traffic.

To implement and establish a sustainable mobility system, certain approaches and principles must be observed. The development of this field facilitated the development of these approaches and principles; a brief summary of the basic principles of sustainable traffic and mobility is provided below:

- *The approach must be comprehensive/holistic*: The approach to the realisation of sustainable mobility must be well planned. For that, we need suitable information and particularly effective analytical tools. A comprehensive analysis must enable decision-makers and all involved stakeholders to understand their subsequent decisions and foresee their effects. The analysis must not be limited, for example, to mere financial effects and market activities.
- *The planning must be strategic and integrated*: When planning sustainable mobility, individual decisions must support long-term strategic goals of the wider community. To this end, the planning of the transport system must be harmonised with environmental, economic, and social plans.
- *We must focus on goals and results*: Sustainable mobility plans must be devised on the basis of a vision, goals (e.g., to meet accessibility

and environmental protection requirements, to enhance social welfare etc.), and the analysis of causes for the existing situation.

- *We must respect equality*: When planning sustainable mobility, the equality effect in our society must be taken into account, which applies to both present and future generations.
- *Observing the precautionary principle*: This emphasises the inclusion of risk in decision making and support for policies. The latter must reduce risks as much as possible.
- *Observing the preservation ethics principle*: When planning sustainable mobility, priority must be given to solutions that preserve our resources, enhance our effectiveness, and reduce resource consumption.
- *Public involvement and transparency principles*: Among the most important principles when planning sustainable mobility is the public involvement and transparency principle. The planning process must be transparent and clearly defined. Equal opportunities of all stakeholders must be ensured both in the field of information and participation in decision-making procedures. Good communication between experts and the interested public is also very important.
- *Forms of mobility are equivalent*: When planning sustainable mobility, the fact that each mode of transport has its advantages and disadvantages from the aspects of the speed of vehicles, capacities, flexibility, energy consumption, road safety, and environmental impacts must be taken into account. When selecting the type of mobility to meet our need for mobility to the greatest extent possible, our decision is based on the advantages and disadvantages of individual modes of transport. In certain cases, it is sensible to take into account a combination of various types of mobility to emphasise their advantages. The so-called "transport chain" may be established on this basis, which is more effective, cost-efficient, and sustainable.

- *Polluter pays principle*: This principle must be observed also when planning sustainable mobility; it is frequently the case in the market that the price of a product (or a service in the case of transport) does not include all costs. The sustainable approach requires the polluter to also pay internalised costs. In the end, a comprehensive observation of this principle requires a market reform; incentives for excessive use of natural resources and (permanent) environmental destruction should be eliminated.
- *'Prevention is better than cure' principle*: Like in certain other fields, sustainable mobility also requires problems to be prevented rather than solved at a later stage. Therefore, the observation of a holistic approach is extremely important.

The aforementioned principles are not all principles according to which sustainable mobility is established/formed. But we may claim that the observation of all the stated principles would contribute to reducing the undesired effects of traffic/mobility.

SUSTAINABLE ROAD SAFETY

Road safety is one of the fundamental qualities of the transport system. Each road user wishes to have a system that meets their needs and expectations. Safe road traffic is also the responsibility of the state whose institutions have an insight into, and an overview of, traffic, as well as the necessary mechanisms to directly or indirectly affect this field in addition to individuals. The quality of life of all citizens depends on the level of road safety.

How to Improve Road Safety?

Safety may be improved:

- by stimulating road users to behave more responsibly;
- by complying with regulations;
- by raising awareness of the significance of road safety;
- by producing safer vehicles; and
- by providing safer road infrastructure.

In a complex system, road safety depends on:

- responsible behaviour of individuals;
- educational and preventive organisations;
- the media;
- repressive and judicial authorities;
- the civil society;
- companies, the leadership of local communities; and
- state authorities.

Visions of Road Safety

The vision of road safety is a description of the desired state based on the theory of how various elements of the traffic system interact (or should interact).

It is defined as a long-term goal without a precisely determined time frame which may be filled only with great efforts over a longer period. Nevertheless, the vision provides concrete guidelines for efforts in the field of road safety, and prompts contemplation of which improvements are necessary to move towards the desired state declared by the vision. Enthusiasm and financial resources can support the vision of road safety in guiding measures, and forming the basis of plans and programmes in this field.

The most renowned examples of visions of road safety are:

- "Sustainable Safety" in the Netherlands; and
- Swedish "Vision Zero."

On the basis of both visions, similar visions were adopted also in other countries worldwide. The main concept of both visions is virtually the same:

- to transform the road transport system into a system that excludes all known opportunities for human error, and
- reduces physical damage in road accidents which may happen.

Since this vision is common to all the interested parties (all stakeholders in the system), road safety is the responsibility of:

- road users;
- system designers;
- road operators;
- car manufacturers etc.

Sustainable Safety

The vision titled "Sustainable Safety" (SWOV, 2006) significantly affects efforts for road safety in practice, and guides towards the implementation of more effective and sustainable measures in this field. The goals of a sustainably safe road system are:

- to prevent road accidents; and
- to reduce their consequences if accidents happen as much as possible.

The vision of Sustainable Safety is based on the idea that people make mistakes and are physically vulnerable. It includes five main principles:

- functionality (of road/road networks);
- homogeneity (of masses and/or speed and direction of road users);
- predictability (of road design and the behaviour of users);
- forgiveness (of road/roadsides and users); and
- awareness (of users).

An example of sustainable and effective measures:

- One of the consequences of, for example, the principle of homogeneity is that motor traffic and vulnerable road users (pedestrians, cyclists) can only interact if the speeds of motorised traffic are low.
- If the speeds cannot remain low, vulnerable road users must be separated from other road users using certain means.
- Measures to realise that include significantly higher numbers and larger sizes of 30 km/h zones in built-up areas, the introduction of 60 km/h zones outside built-up areas, and lower speeds at intersections.

Who is included in “Sustainable Safety” in the Netherlands? Sustainable Safety has been the leading vision of the policy on road safety in the Netherlands since the 1990s. Sustainable Safety measures are carried out by offices of roads at various levels (state, regional, and local). The question of the effectiveness and costs of the establishment of sustainable safety arises. It is estimated (in the Netherlands) that the infrastructural measures of the sustainable safety approach reduced the number of fatalities and hospitalised persons by six per cent across the country. Costs, particularly costs related to road reconstruction, are high, but most of them may be included in the budget for regular maintenance.

The basic fields defined as special priority fields in Dutch Sustainable Safety are:

- Infrastructure
- Vehicles
- ITS
- Education
- Legislation and its implementation
- Special treatment fields include "speed management," driving under the influence of intoxicating substances, young drivers and beginner drivers, pedestrians and cyclists, scooters and motorcycles, heavy goods vehicles.
- The third edition of Sustainable Safety was published in 2018 (SWOV, 2018).
- The five basic concepts/principles of road safety were improved and corrected in accordance with new findings which are the basis of individual specific solutions.
- Three out of five principles refer to planning (design principles):
    1. FUNCTIONALITY of roads;
    2. (BIO)MECHANICS: limiting differences in speed, direction, mass and size, and giving road users appropriate protection;
    3. PSYCHOLOGY: aligning the design of the road traffic environment with road user competencies.
    4. The remaining two principles refer to the organisation of the system (organisation principles):
    5. Effective allocation of RESPONSIBILITY;
    6. LEARNING and INNOVATION in the traffic system.

Concerning design principles, vulnerable modes of transport (pedestrians and cyclists in particular) and the competencies of older road users are now given more attention. The third edition of the Sustainable Safety vision (SWOV, 2016) pays greater attention to cyclist crashes not involving motorised vehicles. Responsibility is emphasised with respect to the role and potential actions of stakeholders in realising an inherently safe road traffic system. The third edition of the Sustainable Safety vision advocates in-depth analysis of all fatal road crashes to learn from the things that still go wrong. Furthermore, this third edition of the vision calls for a pro-active and risk-based approach, using both crash statistics and road safety performance indicators (or surrogate safety measures) as safety indicators and as the basis for action. The aim is to work systematically towards maximum road safety for everybody by means of this third edition of Sustainable Safety, the ultimate ambition being a casualty-free traffic system. In other words, in the end, every road user – be it a schoolchild, a commuter, a commercial driver or an active senior – will come home safely!

Vison Zero

In 1997, the Swedish parliament adopted Vision Zero, a daring policy on road safety based on four principles (Tingvall and Haworth, 2000):

- Ethics: human lives and health are the most important; they take priority over mobility and other goals of the road transport system.
- Chain of responsibility: providers, expert organisations, and professional users are responsible for the safety of the system. The task of users is to comply with regulations. If road users do not comply with regulations, the responsibility is transferred again to system providers.
- Philosophy of safety: people make mistakes; road transport systems should reduce opportunities for errors and damage caused by them as much as possible.
- Stimulating mechanisms of change: providers and operators of the road transport system must strive to ensure the safety of all citizens to the greatest extent possible, and each user must be prepared for changes to achieve such safety.

Who Is Included?

The Swedish Road Administration (SRA) is fully responsible for road safety within the road transport system. In accordance with the principles of Vision Zero, all other interested parties in the field of road traffic must ensure and improve road safety.

Effectiveness and Costs?

It is estimated that Vision Zero might reduce the number of fatalities by one fourth and one third over ten years. The adoption of Vision Zero in Sweden contributed to the development of further research and the implementation of a new design of the system. It contributed to, for example, the reconstruction of single carriageways into 2+1 roads with safety fences in the median to protect drivers from oncoming traffic.

When Vision Zero was first introduced in 1995, it represented a whole new way of viewing problems concerning road safety – including how those problems should be solved. Vision Zero emphasises that the road transport system is an entity in which different components, such as roads, vehicles and road users, must interact in order to ensure safety. Never before has there been this kind of overall perspective. Vision Zero alters the view on responsibility. Entities who design the road transport system bear the ultimate responsibility for safety: road managers, vehicle manufacturers, carriers, politicians, public employees, legislative authorities, and the police. It is the responsibility of everyone to abide by laws and regulations. So far, practically all the responsibility has been put on individual road users. Vision Zero is composed of several basic elements, each of which affects road safety.

These elements concern ethics, human capability and tolerance, responsibility, scientific facts, and the realisation that different components in the road transport system interact and are interdependent.

ROAD INFRASTRUCTURE SAFETY MANAGEMENT

Every year, more than a million people die in road accidents around the world, and about 70 per cent of these deaths occur in developing countries. Pedestrians represent 65 per cent of road accident deaths and 30 per cent of them are children.

In addition, a staggering 20 to 50 million people are injured or disabled each year in road accidents in developing countries, often pedestrians, motorcyclists, cyclists, and non-motorised vehicle occupants.

Almost 85,000 people died in the European Region from road traffic injuries in 2013 – more than 230 every day! Road accidents are the leading cause of death in young people aged 5–29 years. Almost 40 per cent of people who die on the roads are pedestrians, cyclists, and motorcyclists.

For every person who dies in a road accident, at least 23 have non-fatal injuries requiring hospital admissions and many more require emergency room attendance. According to the findings that road safety is a global problem, the European Union defined new action strategies to implement measures for better road safety, operators, and the monitoring of the efficiency of implemented measures. Above all, the strategies refer to the implementation of measures to decrease the number of fatalities which amount to approximately 50,000 every year in the EU.

In 2001, the European Union set the goal to halve the number of deaths on European roads until 2010. Therefore, the European Commission announced in the White Paper (EU, 2001) on European Transport Policy for 2010 to adopt a decision about the implementation of tangible measures in terms of increasing the level of road safety in road infrastructure.

Two of those measures are:

- Directive 2004/54 on minimum safety requirements for tunnels in the Trans-European Road Network;
- Directive 2008/96 on Road Infrastructure Safety Management.

As explained, better (road) road safety can be provided in many ways, i.e., by:

- promoting responsible behaviour of road users;
- adopting repressive measures in terms of fines for irresponsible drivers;
- complying with rules, and raising awareness about the importance of road safety;
- designing safer vehicles; and
- designing, constructing, and maintaining safer road infrastructure.

Directive 2008/96 on Road Infrastructure Safety Management

This Directive requires the establishment and implementation of procedures relating to road safety impact assessments, road safety audits, the management of road network safety, and safety inspections by Member States. It applies to roads which are part of the trans-European (Trans EuropeaN – TEN) road network, whether they are at the design stage, under construction or in operation.

Member States may also apply the provisions of this Directive, as a set of good practices, to national road transport infrastructure not included in the trans-European road network, which was constructed, in whole or in part, with EU funds. Note: the UK, the Netherlands, Belgium etc. have already applied, Slovenia started in 2011.

Safety Audits and Safety Inspection

The number of severe accidents in the EU is unacceptably high. One reason for this situation is that the road safety performance of a part of the existing network is not suitable. Many roads were designed and constructed a few decades ago for less traffic and slower cars. As a result of a basic research project, it is evident that, in every third accident, the road environment has considerable influence.

The benefits of safety audits and safety inspections are in:

- minimising the risk of accidents occurring in the future as a result of planning decisions on new transport infrastructure schemes;
- reducing the risk of accidents occurring in the future as a result of unintended effects of the design of road schemes;
- reducing the long-term costs associated with a planning decision or a road scheme;
- enhancing the awareness of road safety needs among policy-makers and scheme designers.

To be effective, treatments on road infrastructure and road surroundings must be identified and implemented as a result of RSI and RSA. Research by Rune Elvik shows significant expected accident reductions as a result of road safety inspections and associated remedial works.

Examples, for instance, include:

- correcting incorrect signs: 5–10 per cent reduction,
- adding guardrails along embankments: 40–50 per cent reduction
- providing clear recovery zones: 10–40 per cent reduction
- removing obstacles in the visibility area: 0–5 per cent reduction.

Road Safety Audit (RSA)

A road safety audit is a formal procedure for an independent assessment of the accident potential and likely safety performance of a specific design for a road or traffic scheme – whether new construction or an alteration to an existing road.

Road safety audits form an integral part of the design process of the infrastructure project at the stage of:

- draft design;
- detailed design;
- pre-opening; and
- early operation.

In most countries, design guidelines are applied, which only implicitly consider road safety issues and usually reflect a compromise between other road transport issues. For this reason, accidents can also occur on new roads despite the application of effective guidelines, which is due to various factors. First, design guidelines often contain mere minimum requirements regarding road safety. An inauspicious combination of design elements (in horizontal and vertical alignment) with minimum standards can therefore lead to hazardous road stretches. Furthermore, during the design process, a road designer must keep several issues in mind, which affect the design itself.

Road Safety Inspection (RSI)

A road safety inspection (RSI) is a systematic field study, conducted by road safety expert(s), of an existing road or section of road to identify any hazards, faults, and deficiencies that may lead to serious accidents. In short – RSI is for EXISTING roads! Following the principle *"Prevention is better than cure,"* RSI enable the evaluation of existing road traffic facilities, and improve road safety performance.

A road safety inspection is a safety management tool that can be implemented by road authorities as part of an overall road safety management. The RSI goal is to establish potential problems, and propose sufficient countermeasures to reduce the number of accidents or minimise accident severity. This in turn will lead to lower costs associated with accidents, individuals, families, and society.

It is important to note that:

- An RSI is systematic – this means it will be carried out in a methodical way following a formal procedure.
- An RSI is pro-active, trying to prevent accidents through the identification of safety deficiencies for remedial action, rather than react after accidents happen.
- An RSI relates to existing roads, not roads under construction (these constitute the subject of road safety audit).
- An RSI should be carried out by an independent person or team with experience in road safety work, accident analysis, traffic engineering, road user behaviour, and/or road design.
- An RSI is an approved tool to improve the road environment factors.

With expert knowledge of inspection and systematic RSI, it is possible to reduce the number and the severity of road accidents by improving the road safety performance of existing roads.

Road Safety Impact Assessment (RSIA)

An RSIA is the strategic comparative analysis of the impact of different planning alternatives for a new road or a substantial modification to the existing network on the safety performance of the road network.

The primary purpose of a road safety impact assessment (RSIA) is to demonstrate, on a strategic level, the implications on road safety of different planning alternatives of an infrastructure project. A road safety impact assessment is carried out at the initial planning stage of a project, and is continually reviewed through the design phases until scheme approval. Member States ensure that a road safety impact assessment is carried out for all infrastructure projects. A road safety impact assessment indicates the road safety considerations which contribute to the choice of the proposed solution. It further provides all relevant information necessary for a cost-benefit analysis of the different options assessed.

An RSIA may be carried out by an auditor who is a neutral legal person and independent of the designing/planning team. Road safety auditors who carry out functions under Directive 2008/96/EC should undergo initial training resulting in the award of a certificate of competence, and should take part in periodic further training courses.

Network Safety Management (NSM)

While black spot management (BSM) is one of the reactive approaches within the discipline of road safety, network safety management (NSM) is a proactive component. Generally, it is used for planning cycles every two to four years. In NSM, crash data are blended with traffic volumes from traffic counts or ideally from a transport model. Planners and road safety analysts are therefore in a position to assess the most critical points in the network according to their traffic significance, identifying these and creating a priority ranking.

In terms of network safety management, Directive 2008/96/EC is very clear:

- Member States ensure that the ranking of high accident concentration sections and the network safety ranking are carried out on the basis of reviews, at least every three years, of the operation of the road network.
- Member States ensure that road sections showing higher priority according to the results of the ranking of high accident concentration

sections and from network safety ranking are evaluated by expert teams by means of site visits.

- Member States ensure that remedial treatment is targeted at those road sections. Priority is given to measures which present the highest benefit-cost ratio.
- Member States ensure that appropriate signs are in place to warn road users of road infrastructure segments that are undergoing repairs and which may thus jeopardise the safety of road users. These signs also include signs which are visible during both day and night time and set up at a safe distance.
- Member States ensure that road users are informed of the existence of a high accident concentration section by appropriate measures.

The identification of road sections with a high accident concentration takes into account at least the number of fatal accidents that have occurred in previous years per unit of road length in relation to the volume of traffic and, in case of intersections, the number of such accidents per location of intersections.

The identification of sections for analysis in network safety ranking takes into account their potential savings in accident costs. Road sections must be classified into categories. For each category of roads, road sections must be analysed and ranked according to safety-related factors, such as accidents concentration, traffic volume, and traffic typology. For each road category, network safety ranking must result in a priority list of road sections where an improvement of the infrastructure is expected to be highly effective.

Infrastructure Measures for Ensuring Sustainable Safety and Sustainable Traffic

In the following sub-chapters, certain infrastructure measures which aim to ensure sustainable road safety, and sustainable traffic goals are described and discussed.

Self-Explaining Roads

The main problem with the majority of roads is that they have been designed, constructed, and maintained in accordance with traffic and technical requirements, which are implemented in a deficient, inconsistent, and outdated legislation. The impact of proper road and roadside design on safe behaviour of road users has often been neglected.

In Europe, a new (advanced) concept of road design was developed at the beginning of the 21st century as a response to the decline in road safety. Human beings are positioned as the central and the most important factors of road safety together with their limited abilities.

> "In a sustainably safe road traffic system, infrastructure design inherently and drastically reduces crash risk. Should a crash occur, the process that determines crash severity is conditioned in such a way that severe injury is almost excluded." (Koornstra et al., 1992)

The new concept of road design is included in the Dutch concept of sustainable road safety, which places prevention at the forefront, underlining it as more important than curative road safety.

The concepts of self-explaining and forgiving roads are included in the guiding principles of road design in accordance with sustainable safety. (Prestor et al., 2014; Renčelj and Tollazzi, 2016).

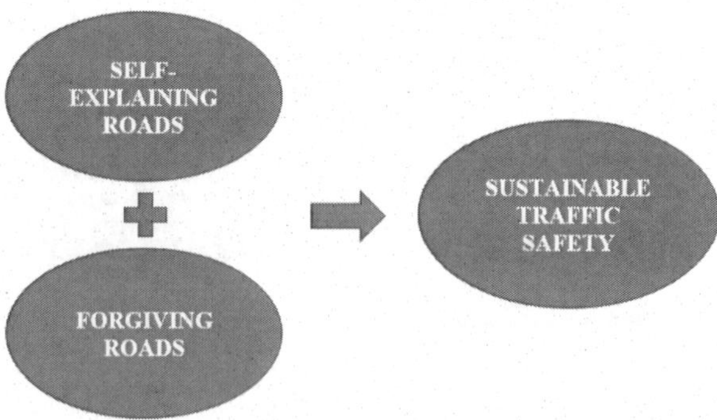

Figure 1. Guiding principles of road design in accordance with sustainable safety.

The Concept of Self-Explaining Roads (SER)

The beginners of SER are Theeuwes and Godthelp, who in 1992 published an article titled "Begrijpelijkheid van de weg," which means "understandable roads" in Dutch. The authors used the English term "self – explaining roads" because they believed that the term "understandable roads" did not appropriately describe complex mental processes. (Theeuwes and Godthelp, 1992).

The SER concept has spread across the world (the Netherlands, Denmark, Germany, Great Britain, Australia, New Zealand etc.). By definition, SER are roads which only by their form induce safe behaviour of all road users (SPACE, 2012). A SER is a road designed and built so as to induce adequate behaviour and avoid driving error. A perfectly designed SER would not require speed limit signs or any warning signs! (FHWA, 2001).

Key Terms and Goals of SER

The key terms in the SER concept are (a) categorisation and (b) perception, and consequently, the expectations of the road users. The goal of the SER is to design the road environment that is aligned with expectations. An interaction between appropriate drivers' expectations and the road environment constitutes traffic atmosphere, which is a condition for safe behaviour. SER connect the categorisation of road networks and expectations of road users.

The traffic environment induces the right expectations of road users regarding the presence and behaviour of other road users, as well as regarding their own behaviour (Martens et al., 1997). To attain this goal, clearly separated categories of roads should be implemented, whereby each road category should clearly define particular behaviour of all road users.

Conditions for Characteristic Road Categories

The characteristic road categories system should meet the following conditions (SPACE, 2010):

- each category should consist of unique road elements (homogeneous within one category and different from all other categories);
- each category should require unique behaviour for a specific category (homogeneous within one category and different from all other categories);
- unique behaviour displayed on roads should be linked to unique road elements;
- the layout of crossings, road sections, and curves should be linked uniquely with the particular road category;
- one should choose road categories that are behaviourally relevant;
- the same road category should connect the road section, which is psychologically interpreted as a whole;
- there should be no fast transitions going from one road category to the next;
- when there is a transition in road category, the change should be marked clearly;
- when teaching the different road categories, one should not only teach the name of, but also the behaviour required for, that type of road;
- category-defining properties should be visible at night as well as in the day-time;
- the road design should reduce speed differences and differences in direction of movement;
- road elements, marking, and signing should fulfil the standard visibility criteria;
- the traffic management systems should be clearly connected with special road categories.

The establishment of the SER network with the goal of reducing unintentional incidents must be accompanied with systematic and interdisciplinary measures concerning infrastructure and education of users.

Requirements and Elements of SER

Requirements for SER:

- Clear and consistent Functional Categorisation
- Road Layout:
  - Recognisable
  - Distinguishable
  - Easily interpretable
- Limited number of road categories (3–4)
- Traffic regulation adapted to road categories (Traffic mix, Overtaking, Speed)
- Transitions:
  - At intersections
  - At town/settlement entrances.

The following features could be relevant for promoting the recognition of SER types:

- Longitudinal markings
- Driving direction separation
- Lane width
- Adjacent cycle lanes
- Road surface/extent of roughness
- Characteristics of shoulders (width, obstacle distance, reflector posts)
- Roadside environment (land use, for instance: urban characteristics such as
- buildings, parked cars, exits)
- Intersections and transitions type (not continuously visible).

Traffic Calming

Traffic calming is generally a combination of different measures to reduce the negative influence of motorised traffic, change the behaviour of motor vehicle drivers, and improve traffic conditions for non-motorised road users. The main goal of traffic calming should be to improve road safety for pedestrians and cyclists without severely limiting vehicle travel or significantly effecting emergency vehicles. The purpose of traffic calming is not to block traffic, but to slow it down.

Especially in the last twenty years, traffic calming measures (all kind of devices, applications etc.) have become more and more frequent in Slovenia. The Slovenian act on public roads defines measures and devices for traffic calming. According to our law, traffic calming devices are physical, light or other devices and obstructions that physically prevent road users from driving at inappropriate speed or warn them to limit the speed on dangerous road sections.

Physical obstructions in Slovenia could (only) be placed on regional roads and community (local) roads in settlements/cities. Physical traffic calming devices must be used near schools, kindergartens, and other buildings/facilities along which the speed is limited (due to road safety).

One of the main goals of adopting traffic calming measures – especially installing physical devices – is to reduce the speed of motor vehicles. Lower speed usually results in fewer road accidents; the number of road accidents and their consequences, especially when a vehicle/pedestrian (cyclist) is involved, lowers.

Why Do We Need Traffic Calming?

According to NTF – the Swedish National Society for Road Safety policy (NTF, 2006) we could say, that if the average speed on a road is changed by x per cent, the number of accidents changes by twice x per cent, the number of injured by three times x per cent, and the number of people killed by four times x per cent. Therefore, if we reduce the average speed by 10 per cent, we could say that we will have 20 per cent fewer road accidents in which there will be 30 per cent fewer injured people and 40 per cent fewer fatalities.

We could say something similar about traffic calming measures: the more we reduce the speed of motor vehicles, the better results we could get from the road safety point of view. But this does not always work; there can be exceptions.

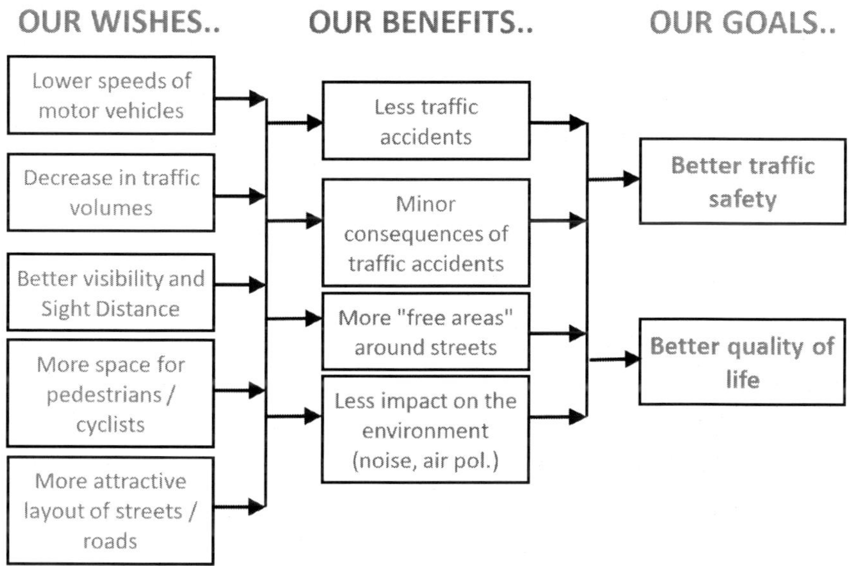

Figure 2. What we would like, and what our benefits and our goals are in traffic calming.

Types of Devices and Physical Traffic Calming Measures

In general, we could define 4 different types of traffic calming measures:

- Systemic measures (measures in the road network)
- Traffic signs (limitations etc.)
- "Warning devices" (optical breaks etc.)
- Physical traffic calming measures.

Inside urban areas, we could establish traffic calming measures according to locations:

- At the beginning of settlements
- On road sections
- At intersections.

Figure 3. Traffic calming measures in urban areas/settlements: at the beginning of settlements, on road sections and, at intersections.

Traffic Calming Measures at the Beginning of Settlements

"Optical breaks" (also known as "optical obstacles" / rumble strips) (LCC, 2006) are so-called "mild" measures, slightly raised strips of different coloured surfacing (normally white) set across the width of the driving lane.

The appearance and feel of the strips are intended to make drivers reduce their speed. Normally they are used before areas with speed limits (for example, at the beginning of settlements) and are placed in non-equal distances perpendicular to the driving direction.

Central islands (mid-block medians) are traffic calming measures. They are raised traffic islands which are located along the centreline of a road or street (Figures 5, 6). They may be combined with lane narrowing. Central islands are sometimes called mid-block medians, median slow points or

Figure 4. Optical breaks.

median chokers. Central islands on the roads and streets of cities (settlements etc.) are often landscaped. They provide a visual amenity and support the neighbourhood identity. As one of the important traffic calming measures, central islands could also help to make our roads and streets more "pedestrian-friendly"; they work very well when they are combined with pedestrian crossings (a central island is a mid-point refuge for pedestrian crossings). As mentioned before, central islands could also be used on wide streets to narrow travel lines. On Slovenian roads and streets, central islands are used as traffic calming measures particularly:

- at the beginning of cities/settlements; drivers are warned that they should reduce their speed;
- in cities/settlements; pedestrian protection at pedestrian crossings prevents prohibited vehicle manoeuvres.

The advantage of central islands is that, if they are well designed, such central island narrowing increases pedestrian safety. They can also have positive aesthetic value and reduce traffic volume. The disadvantage is that the speed-reduction effect is somewhat limited, and they may require the elimination of on-street parking. There has also been some specific research about central islands as traffic calming measures in Slovenia (Renčelj, 2004; 2005).

Figure 5. Central island.

Figure 6. Transition zone (at the entrance to an urban area) equipped with a central island.

Figure 7. Re-arrangement of the road/different road elements at the beginning of settlements.

Traffic Calming Measures on Road Sections in Settlements

On road sections in settlements, different traffic calming measures may be used, which is shown and described below.

Speed humps/Speed bumps are round raised areas placed across the roadway. The profile of a speed hump can be circular, parabolic, or sinusoidal. Speed humps are good for locations where very low speed is desired and reasonable, and noise and fumes are not of major concern.

The advantages of speed humps are that they are relatively inexpensive, and very effective in speed reduction. However, they have several disadvantages:

- they cause a "rough ride" for all drivers;
- they force large vehicles, such as emergency vehicles and those with rigid suspensions, to travel at slower speeds;
- they may increase noise and air pollution; and
- their aesthetics is questionable.

Figure 8. Speed humps/bumps.

Figure 9. Speed platform.

Figure 10. Roadway narrowing.

Figure 11. Central island with a pedestrian crossing at the location of a bus stop.

Trapezoidal humps/Speed platforms/Speed tables are flat-topped speed humps often made of brick/stone or other textured materials on the flat section. Speed tables are usually long enough to accommodate the entire pedestrian crossing. Their long flat fields give them higher design speeds than speed humps. Speed tables are also good for locations where low speed is desired, but a somewhat smooth ride is needed. The advantages of speed platforms are:

- they are smoother than speed humps; and
- the reduce speed effectively.

Roadway narrowing is not yet (currently) widely spread on Slovenian roads. It could be done from one side or from both sides of the road.

Traffic Calming Measures at Intersections in Settlements

A realigned/modified intersection means changes in alignment, often converted T-intersections with straight approaches into curving streets. They are one of few traffic calming measures for T-intersections. Realigned intersections can be effective at reducing speeds and improving safety at T-intersections.

Raised intersections are flat raised areas covering the entire intersection, with ramps on all approaches, and often made of brick or other textured materials on the flat section. They are usually raised to the level of the pavement. By modifying the level of an intersection, pedestrian crossings are more readily perceived by motorists to be "pedestrian territories." Raised intersections are good for areas where other traffic calming measures are unacceptable. The advantages of raised intersections are:

- they improve safety for pedestrians and vehicles;
- they can have positive aesthetic value and
- they can calm two streets at the same time.

Small roundabouts (one-lane roundabouts, mini-roundabouts) are usually located:

- in areas with a history of accidents;
- at intersections where queues need to be minimised;
- at intersections with irregular approach geometry, where there is a high proportion of U-turns; and
- on locations with abundant right-of-way.

Figure 12. Realigned/modified intersection.

Figure 13. Raised intersection.

Roundabouts can moderate traffic speed on arteries, are generally aesthetically pleasing, enhance safety compared to traffic signals, can minimise queuing at the approaches to the intersection, and are less expensive to operate than traffic signals. These are the advantages of roundabouts. On the other hand, they may be difficult for large vehicles, and must be designed in a way that the circulating lane does not encroach on pedestrian crossings. They may require the elimination of on-street parking.

Figure 14. Small one-lane roundabout.

Effectiveness of Traffic Calming Measures – Slovenian Experiences

We have already observed the effectiveness of traffic calming measures on Slovenian roads. For example, in 2005, a wide research at the national level (Martinčič et al., 2005) was performed. We could point out the effect of different traffic calming measures on driving speeds (the results have already been published (Renčelj, 2007). One of the goals of the said research, which was done in Slovenia (from 2003 to 2005), was also to define the real effect of speed reduction on different types of traffic calming measures across Slovenia. In the research, we selected 32 typical locations in Slovenia, where six different types of traffic calming measures appear. The main goal of this part of the research was to establish the effectiveness of different types of traffic calming measures. For this purpose, hidden speed measurements with laser measurement instrument Riegl LR90-235/P were used. At that time, we also performed measurements of other dimensions (e.g., precise dimensions of traffic calming devices, the dimensions of road elements and their surroundings, traffic counting, a questionnaire etc.). The main findings of our research are presented in Table 1.

Table 1. Results of hidden speed measurements at different types of traffic calming measures

Type of traffic calming measure	Speed humps	Speed platforms	Raised intersections	Optical breaks	Central islands	One-lane roundabouts
Number of observed locations	3	5	4	6	5	3
Number of measurements	166	344	352	516	446	367
Vmax [km/h]	35–39	28–44	29–43	59–93	61–96	45–50
Vaver [km/h]	16.8–24,5	15–22	16.2–22	42.5 – 60.6	43.7–66.1	24–26.9
V85 [km/h]	21–30.5	18.5–25.1	20.4–27.5	50.4–70.7	51.1–81.8	31.8–34.4
V85 aver [km/h]	27.03	22.6	23.4	58.43	60.9	32.77

Mini-Roundabouts

Mini-roundabouts are small roundabouts with a fully traversable central island. They are most commonly used in low-speed urban environments with average operating speeds of 50 km/h or less. They can be useful in such environments where conventional roundabout design is precluded by right-of-way constraints. In retrofit applications, mini-roundabouts are relatively inexpensive because they typically require minimal additional pavements at the intersecting roads and minor widening at the corner curbs. They are mostly recommended when there is insufficient right-of-way to accommodate the design vehicle with a traditional single-lane roundabout. Because they are small, mini-roundabouts are perceived as pedestrian-friendly with short crossing distances and very low vehicle speeds on approaches and exits. A fully traversable central island is provided to accommodate large vehicles and serves one of the distinguishing features of a mini-roundabout. The mini-roundabout is designed to accommodate passenger cars without requiring them to traverse over the central island. The overall design of a mini-roundabout should align vehicles at entry to guide drivers to the intended path and minimize running over of the central island to the extent possible (Rodegerdts et al., 2010).

In Slovenia, we implemented the first mini-roundabout in 2002. Since then, we have built more than 50 mini-roundabouts across the country.

Proper Design Parameters of Mini-Roundabouts

When deciding to construct a mini-roundabout, it is necessary to consider the characteristics of the existing road network in the surrounding area (e.g., existing types/designs of intersections etc.), the existing ways of traffic routing, and the "expectations of users," i.e., road users.

Mini-roundabouts can be built instead of existing "classic" three- and four-lane intersections, which can lead to a reduction in the number of road accidents, a reduction in delays and queues at the intersection, as well as reduction in the speeds of motor vehicles (as an individual traffic calming measure or combined with other traffic calming measures).

Mini-roundabouts can be built only on those roads in settlements where the maximum allowed speed is 50 km/h (or lower). In addition, the measured speed V_{85} on approach roads to a mini-roundabout (within the distance of 70 m from the mini-roundabout) is lower than 50 km/h. If the measured speed V_{85} on approach roads exceeds 50 km/h, a mini-roundabout must be built together with other traffic calming devices and measures.

A mini-roundabout is a good solution when reconstructing existing intersections (and as a corrective measure) in different environments – built-up areas, residential areas, business areas or shopping areas. Mini-roundabouts are also suitable when reconstructing existing intersections:

- intersections of irregular shapes, such as shapes of letters Y, K, A and X,
- intersections in the shape of letters F and H (two consecutive T-intersections within a short distance);
- when traffic loads are approximately the same on main and minor traffic routes;
- where the installation of traffic-lights is not justified, but the capacity of an intersection with no traffic-lights is exceeded;
- when the main traffic route is unsuitable in relation to the existing intersection geometry.

Mini-roundabouts are not the most suitable solution when the share of large motor vehicles on the main route is high (freight vehicles and/or buses). The construction of mini-roundabouts is not recommended on significant public transport lines, in industrial/manufacturing zones etc.

Possible Problems with Mini-Roundabouts

Although mini-roundabouts –are generally a solution with a positive effect on road safety, we could detect some possible problems/disadvantages of mini-roundabouts. For example:

- unsuitable splitter island construction (unsuitable design, construction – in correlation with posted traffic signs

- road markings (visibility and maintenance of road markings);
- visibility of the central island (not just the central island painted with road markings; a better solution is a mountable cobblestone island; the edge de-levelled by 2–3 cm);

Figure 15. Mini-roundabout.

- unsuitable width of driving lanes (before the mini-roundabout or inadequate widening of driving lanes before entering the-mini roundabout);
- unsuitable design and construction (in some cases, we a notice lack of knowledge of mini-roundabout design and construction);
- traffic rule violations (it is in connection with inadequate mini-roundabout design and construction, especially splitter island construction).

Examples of an Interdisciplinary Approach/Search for Solutions at the Level of Student Tasks and Master's Theses

This chapter includes a brief description and presentation of some master`s theses, whose topic refers to the content of this chapter.

Self-Explaining Roads

This master's thesis (Prestor, 2014) discusses the problems of the concept of self - explaining roads, which, together with the concept of forgiving roads, is the basis of safe road design. The master's thesis describes the development and comparison of the SER roads concept in Europe, and emphasises the problem of roads with several traffic functions. Based on the analysis of the classification of the road system in the Netherlands, Denmark, and Germany, the analysis of the basic conditions for the establishment of the SER road system, and on the analysis of the existing classification of the road system in Slovenia, an optimal SER classification of the road system in Slovenia is introduced.

Error Forgiving Roads

The next master thesis (Stojan, 2016) discusses the issue of safe road and roadside design on secondary roads outside settlements. The basic principles, objective and purpose of the "forgiving roads" concept, as well as experience of their use in the several European countries has been studied. For the purpose of the master's thesis, a research/questionnaire was carried out by using the Delphi method. Its purpose was to determine the influence of various technical measures of the concept of forgiving roads on better road safety or safe road design, and what their potential applications/uses in Slovenia would be. Based on the opinions of experts, we have prepared and presented proposals for technical measures of the concept of forgiving roads on the secondary road network (outside settlements) in Slovenia.

Traffic Calming

In the field of traffic calming, many different diplomas works and master's theses were conducted. Those papers deal with different aspects of this topic. For example, traffic calming measures at the level intersections and access points (Njenjić, 2008), types of appropriate traffic calming measures on state roads in the vicinity of schools (Božič, 2009), the laid option of traffic calming measures in the wider area of residential parts of a city (Ramovš, 2009), an analysis of the efficiency of traffic calming measures (Skrivarnik, 2011), hidden speed measurements on the road section – between traffic calming measures, driving speeds over trapezoidal speed tables (Bernad, 2015).

Another aspect is the analysis and proposed implementation of 30 km/h zones in settlements. For example, the proposed implementation for the Municipality of Miklavž na Dravskem polju (Polc, 2013) and road safety analysis in different 30 km/h zones (Fijačko, 2015).

CONCLUSION

Today, sustainable traffic is in direct connection with the living environment – this is also directly related to the interdisciplinary approach in the field of higher education in the area of sustainable building design.

Sustainable mobility is an established approach with which we could reduce the negative effects of road traffic – it does not imply just a restriction of the use of cars, but it encourages and gives people the opportunity to choose (more often) other modes of transport –public transport, bicycles, walking etc.

This chapter includes an overview of the described content. All important topics within this thematic field are discussed and analysed. An important part is the field of the living environment in connection with mobility – which also includes the following key steps: sustainable mobility, sustainable road safety, (road) infrastructure measures to ensure sustainable mobility. The final section presents examples of an interdisciplinary approach/search for solutions at the level of student tasks and master's theses.

REFERENCES

Bernad, D. (2015). *Analiza ukrepov za umirjanje prometa - primer trapeznih ploščadi, Magistrsko delo, Univerza v Mariboru.* [*Analysis of traffic calming measure - example of the speed tables*, Master thesis, University of Maribor].

Božič, T. (2009). *Naprave in ukrepi za umirjanje prometa v okolici šol*, Diplomsko delo, Univerza v Mariboru. [*Traffic calming devices and measures in school surroundings*, Diploma work, University of Maribor].

EU. (2001). *Commission of the European Communities, White paper: European transport policy for 2010: time to decide*, Brussels.

EU. (2016). *Communication from the Commission to the Council and the European Parliament on Thematic Strategy on the Urban Environment*, EU commission.

FHWA: Brewer, J, German, J, Krammes, R, Movassaghi, K, Okamoto, J, Otto, S, Ruff, W, Sillan, S, Stamatiadis, N., Walters, R. (2001). *Geometric Design Practices for European Roads*. McClean, Vancouver.

Fijačko, A. (2015). *Prometno varnostna analiza območij omejene hitrosti (cona 30 km/h)*, Magistrsko delo, Univerza v Mariboru. [*Traffic safety analysis of 30 km/h zones*, Master thesis, University of Maribor].

Koornstra, M. J., Mathijssen, M. P. M., Mulder, J. A. G, Roszbach, R., Wegman, F. C. M. (1992), *Naar een duurzaam veilig wegverkeer: Nationale Verkeersveiligheidsverkenning voor de jaren 1990/2010* [*Towards sustainable safe road traffic: National Road Safety Survey for 1990/2010*], Stichting Wetenschappelijk Onderzoek Verkeersveiligheid SWOV, Leidschendam.

LCC - Leicestershire County Council. (2006). *Road Safety Measure*. [online] Available at: http://www.leics.gov.uk/index/highways/road_pathway_maintenance/road_schemes/road_safety_gallery/gallery_part_4.htm [Accessed 13 July 2008].

Lee, R., Bansen, J., Tiesler, C., Knudsen, J., Myers, E., Johnson, M., Moule, M., Persaud, B., Lyon, C., Hallmark, S., Isebrands, H., Crown, R. B.,

Guichet, B., O'Brien, A. (2010). *Roundabouts: An Informational Guide. Second Edition.* Transportation Research Board.

Martens, M., Comte, S., Kaptein, N. (1997). *The effects of road design on speed behaviour: a literature review*, TNO Report TM 97 B021, TNO Human Factors Research Institute, Soesterberg.

Martinčič, M., Andrejčič Mušič, P., Špolar, A., Križ, A., Šoba, F., Sajovic, J., Brank, L., Lah, J., Likar, B., Renčelj, M., Tollazzi, T., Šibenik, T. (2005). Raziskava učinkovitosti vpliva uvajanja ukrepov za umirjanje prometa na cestah, Raziskovalni projekt, DDC d. o. o. in Univerza v Mariboru. [*Research about the effectiveness of the impact of the introduction of road safety measures, Research project*, DDC Ltd. and University of Maribor].

Njenjić, S. (2008). Preučitev možnosti izvedbe ukrepov za umirjanje prometa v nivojskih križiščih, Diplomsko delo, Univerza v Mariboru. [*Consideration of options for implementing traffic calming measures at level intersections*, Diploma work, University of Maribor].

NTF - The Swedish National Society for Road Safety. (2006). *What does NTF think?*, Policy.

Polc, K. (2013). Uvajanje območij omejene hitrosti v naselju Miklavž na Dravskem polju, Diplomsko delo, Univerza v Mariboru. [*Introduction of limited speed zones in the vilage* Miklavž na Dravskem polju, Diploma work, University of Maribor].

Prestor, J. (2014). Predvidljive ceste, Magistrsko delo, Univerza v Mariboru. [*Self explaining roads*, Master thesis, University of Maribor].

Prestor, J., Klemen, B., Zoltar, S., Renčelj, M. (2014). "Self-explaining roads: Concept analysis and a proposal for the estabilishment in Slovenia." *Proceedings*, Novi Sad.

Ramovš, T. (2009). Celostni pristop k umirjanju prometa - primer Frankovega naselja v občini Škofja Loka, Diplomsko delo, Univerza v Mariboru. [*Integrated aproach to regulation of the traffic - case study of settlement* Frankovo in municipality Škofja Loka, Diploma work, University of Maribor].

Renčelj, M. (2004). Primerjalna analiza sredinskih otokov kot ukrepov za umirjanje prometa na zmanjšanje hitrosti motornih vozil. Varnost v prometu, varnost in zdravje pri delu, Ljubljana. [*Comparative analysis about speed reduction on the central islands - as the traffic calming measure*].

Renčelj, M. (2005). Influence of speed reducing provision (central island) at the beginning of the settlement on vehicle speed. *Suvremeni promet* 25 (1/2): p. 234-237.

Renčelj, M. (2007). Comparative analysis about speed reduction on the different types of the traffic calming measures in Slovenia. *14th International Conference Road safety on four continents,* Bangkok, Thailand, 14-16 November 2007.

Renčelj, M., Tollazzi T. (2016). Advanced concepts of road infrastructure planning and design: self explaining roads and error forgiving roads. *Managing the road traffic safety in the Southeast Europe - Challenges and directions.*

Skrivarnik, J. (2011). Analiza naprav in ukrepov za umirjanje prometa na območju mestne občine Slovenj Gradec, Diplomsko delo, Univerza v Mariboru. [*Analysis of measures for traffic calming in the Slovenj Gradec*, Diploma work, University of Maribor].

SPACE (EU project) - Speed Adaption Control by Self – Explaining Roads. (2011). *Deliverable Nr 4 – Consistent treatment in relation to the severity of a curve*, a driving simulator study, ERA-NET Road.

SPACE (EU project) - *Speed Adaption Control by Self – Explaining Roads*. (2012). Final Report, Deliverable Nr 6, ERA-NET Road.

SPACE (EU project) - Speed Adaption Control by Self – Explaining Roads. (2010). *Deliverable Nr 1 – Self – Explaining Roads Literature Review and Treatment Information*, ERA NET Road.

Stojan, K. (2016). Metodologija za načrtovanje prometno varne cestne infrastrukture, Magistrsko delo, Univerza v Mariboru. [*Methodology for design safe road infrastructure*, Master thesis, University of Maribor].

SWOV. (2006). Advancing Sustainable Safety, F. Wegman, L. Aarts (Eds.), *National Road Safety Outlook for 2005–2020*, Leidschendam.

SWOV. (2018). *Sustainable Safety 3rd edition – The advanced vision for 2018-2030: Principles for design and organization of a casualty-free road traffic system*, The Hague.

Theeuwes, J., Godthelp, H. (1992). *Begrijpelijkheid van de weg (Self-explaining roads), Report IZF 1992 C-8*, TNO Institute for Perception, Soesterberg.

Tingvall, C., Haworth, N. (2000). Vision Zero: An ethical approach to safety and mobility. In: Jraiw, K. (ed). *Road safety & traffic enforcement: beyond 2000: 6th ITE International Conference Road Safety & Traffic Enforcement: Beyond 2000*, Melbourne, 6-7 September 1999.

About the Author

Marko Renčelj, PhD

University Associate Professor

University of Maribor, Faculty of Civil Engineering, Transportation Engineering and Architecture (UM FGPA), Maribor, Slovenia

Marko Renčelj is an associate professor of civil engineering at the University of Maribor, Faculty of Civil Engineering, Transportation Engineering and Architecture (UM FGPA). He obtained his PhD degree at the University of Trieste, Italy (Universita´ degli studi di Trieste, Italia) in 2009. In twenty years of experience in the research and teaching of road infrastructure design, construction and maintenance and also road traffic safety, he obtained valuable leadership skills, considerable experience in project management, and a sense of work organisation – by organising the teaching process, work in the context of international and national research, and applied projects within the Centre for Traffic Infrastructure under the leadership of Prof. Tomaž Tollazzi. His research work refers to road infrastructure, traffic safety effect of road infrastructure, and infrastructure for cyclists. He has been involved in several international and domestic research projects. He has also obtained a license for responsible project design and responsible management of works (by the Slovenian Chamber of Engineers, Section of Civil Engineers) and licence for a traffic safety auditor (by the Slovenian Traffic Safety Agency). He has published several articles in international scientific journals and presented his research at international conferences on intersections' design, road safety and traffic calming.

In: An Interdisciplinary Approach … ISBN: 978-1-53617-302-4
Editors: V. S. Klemenčič et al.

Chapter 4

BUILDING ENERGY MODELLING IN SUSTAINABLE DESIGN PROCESS

Maja Žigart*

University of Maribor, Faculty of Civil Engineering, Transportation Engineering and Architecture, Maribor, Slovenia

ABSTRACT

Building delivery is a complex and holistic process; therefore, the inclusion of sustainable aspects should be achieved during earlier phases of design. New digital tools for planning and assessment can facilitate optimisation of building design and collaboration of multidisciplinary team members.

During the educational process, students are expected to become familiar with various sustainable aspects, including social behaviour, cultural aspects, economically feasible design, flexible design with adaptive use of buildings, active and passive energy concepts, renewable energy use, recycling and waste management, living comfort, water conservation, and other sustainable design principles. This chapter will present especially the inclusion of energy-efficient building design with

* Corresponding Author's Email: maja.zigart@um.si.

integrated use of new digital technologies, such as building information modelling (BIM) and BIM-based building energy modelling (BEM). The integration of building energy modelling into the sustainable design process will be presented on the case of a student workshop at an interdisciplinary master course. Findings and challenges are presented with an emphasis on encouraging interdisciplinary collaboration between students.

Keywords: sustainable design process, BIM, BEM, interdisciplinary, education

INTRODUCTION

The design process is the process of envisioning and creating buildings (in architecture), objects, services or other products with focus on users, solutions for people, physical items or other systems to meet a certain need or solve a problem. But when is a design sustainable and how can we include sustainable development in the design process?

In 1987, the World Commission on Environment and Development (WCED 1987) published a report entitled "Our common future," which came to be known as the Brundtland Report. It was the first to define now common term sustainable development as "*development that meets the needs of the present without compromising the ability of future generations to meet their own needs*." The term sustainable has development evolved in recent years by understanding that it is a complex and holistic approach very difficult to be defined. Bagheri and Hjorth state that sustainability is not defined as an end state, but an evolving ideal of development efforts with no end known in advance. Therefore, planning for sustainable development should change and focus on processes and learning as the main purpose of planning, instead of on fixed goals (Bagheri and Hjorth 2007).

Sustainability and design are closely linked. Braungart and McDonough suggest "*Human beings don't have a pollution problem; they have a design problem. If humans were to devise products, tools, furniture, homes, factories, and cities more intelligently from the start, they wouldn't even need to think in terms of waste, or contamination, or scarcity. Good design would allow for abundance, endless reuse, and pleasure.*" (Braungart and McDonough 2013). Decisions made during the design process impact sustainable development every day by providing for the needs of future generations.

Educating about the sustainable design process is therefore a relevant topic for many design-oriented schools, such as textile design, graphic design, product design, software design, industrial design, and building design. The building sector is also one of the most influential sectors regarding environmental impacts, and abilities to provide social and sustainable development. Architects can do a lot to eliminate negative environmental impacts by adopting sustainable design practices. Linking sustainability and design can be achieved by integrating sustainable thinking into design curricula in higher education by making sustainability the starting point for any exploration (Dewberry and Fletcher 2001).

"*A design concept is created when, through the designer's powers of synthesis, a recognized need and technical capability as represented by the state of the art are matched. When the designer can arrange technical art into useful combinations which form a system satisfying a need, he has a design concept.*" (McCrory 1966) As a sustainable practice, the systems should be applied according to sustainable principles regarding the topics of economic development, social development and environmental protection, satisfying the need of future generations.

It is evident that digital media has been fundamentally changing the way we design, practice, and produce architecture (Al-Qawasmi and Hadjri 2007). Technology has been changing the practice and teaching of architecture. Moreover, with advanced digital applications and computer software, the design process has also been evolving. Analytics and additional data gathered with the help of digital models allow for innovations and optimised performance of sustainable buildings. With mathematical algorithms, 3D printing, virtual and augmented realities, innovative prefabrication processes, and other technological advances, the role of architects has changed and the educational process should adapt to the changes. In design education, it is important to prepare students for the realities of being a designer. On the one hand, they need to acquire design skills and develop individual design identity and on the other hand, they are required to understand and be able to adapt in the professional setting (Leerberg et al. 2010). Future architects should be able to respond to the challenges and opportunities the digital era represents. Digital technologies can provide insight and possibilities for experimentation to achieve building design that would better suit sustainable needs. The inclusion of digital technologies and tools in the architectural education has been researched a lot.

Hasanpour Loumer (2015) emphasises the importance of including digital architecture and intelligent building design as components of sustainable development, because they play a significant role in reducing environmental pollution and climate changes, and providing a healthy life to humans. To include various aspects of sustainable development in building design, we should comprehensively merge new digital technologies with architecture to achieve a broader view on sustainability issues. Architecture schools, like architects, must work with technology and must work to make it more useful. They should be leaders, not merely consumers, in developing design practice with digital media and anticipate, not just adopt, technological change (Gross and Do 1999). There is always a fine line between traditional and contemporary techniques in architecture pedagogy, and the question of when and how digital technologies should be incorporated in the educational process. Kara (2014) argues that the conventional tools of hand drawing and physical modelling should be embraced at the foundation levels of architectural education, whereas digital tools should be introduced after developing a certain set of architectural skills as a sense of tectonic resolution, scale, and spatial experience.

A study from 2007 (Angulo and Válasquez de Velasco 2007) presents first integrations of digital media and multidisciplinary content in three design studios: an electronic design studio, a virtual design studio, and a fabrication design studio at Texas A&M University. The paper ends by addressing the issue of how Building Information Modelling (BIM) can be a pedagogical tool for multidisciplinary integration and academia should provide training for its implementation. In the past decade, BIM has become widely integrated in the industry and the curricula of various architecture and civil engineering schools.

The building information modelling (BIM) concept originated from Professor Charles Eastman at the Georgia Tech School of Architecture in the late 1970s. However, the architecture, engineering and construction (AEC) industry started practically using BIM in construction projects in the mid-2000s. (Ahmad Latiffi et al. 2014) Since it has been implemented in the AEC industry, new perspectives for the use of BIM technology have been introduced, from planning and design, estimation, the construction process, building life cycle, performance, and technology. Based on the BIM technology, a novel approach of Building Energy Modelling (BEM) has emerged, providing an opportunity for the time-efficient, economical, practical, consistent, and accurate building design process with integrated energy performance simulation (Gao et al. 2019).

This article discusses the BIM-based building energy modelling in the sustainable design process with the integration of an interdisciplinary approach in practice and teaching. The following sections of the article discuss the background of building information modelling, the development of BIM-based BEM, and its application in the educational process. On the case of an interdisciplinary student workshop integrated in the educational process, the experience and challenges of application are presented.

BIM in Sustainable Design Process

BIM is a multi-dimensional tool for building life cycle management, also called nD modelling. It can be classified into 3D BIM with a three-dimensional model of building as an upgrade from 2D CAD modelling, 4D BIM used for construction site planning and time-schedule simulations, 5D BIM used for cost planning and analysis, 6D BIM addressing sustainability with thermal analysis and environmental assessment with building certification, and 7D BIM including facility management, building operation and maintenance. (Kushwaha 2016) Additionally, new dimensions in nD models and BIM can be added. To support accident prevention through design, the 8D BIM has been addressed (Kamardeen 2010).

BIM is not just a technology; it may also be viewed as a virtual process that includes different aspects, disciplines, and systems of a building within a single, virtual model. This helps all team members (clients, architects, engineers, contractors, and suppliers) to collaborate more accurately and efficiently than in the traditional design process (Azhar et al. 2012).

The conventional design process mainly has a linear structure because of time restrictions and various team members involved (Larsson 2009). Traditionally, after 2D drawings are created by architects, many other design members, such as civil, structural and mechanical engineers, energy consultants, transportation engineers, and others collaborate to analyse and create solutions based on the design. The architect's design is improved and handed over to experts to achieve efficient and optimised design. Since this process is very time, cost and labour-consuming, the design team usually does not have the possibility to evaluate various design alternatives at earlier design stages.

In the building delivery process, the early design and preconstruction phases are most crucial to making decisions on its sustainability features. Traditional computer-aided design (CAD) planning environments usually lack the capacity to perform sustainability analysis at early design stages. (Azhar et al. 2011) BIM allows designers to easily predict the performance of projects before they are built, optimise designs with simulations, analyses, and visualization, respond to design changes faster, and deliver higher quality building documentation. It also enables extended teams to extract valuable data from the model to help with earlier decision making, and more economical project delivery and facility management. (Kamardeen 2010) BIM facilitates an iterative design and similarly to other industries, helps with the implementation of multidisciplinary design optimisation methods (Sakikhales and Stravoravdis 2015).

BIM working party strategy paper (A report for the Government Construction Client Group Building Information Modelling (BIM) Working Party Strategy Paper 2011) defined BIM maturity model referring to various levels of competence, and the ability to operate and exchange information.

Level 0 represents unmanaged CAD, usually 2D, with paper or electronic paper as the most likely data exchange mechanism. Level 1 is managed CAD in 2D or 3D format with a collaboration tool providing a common data environment, possibly some standard data structures and formats. Level 2 with managed 3D environment is held in separate discipline BIM tools with attached data. Level 3 is a fully open process and data integration enabled by web services, compliant with the emerging IFC standards, and managed by a collaborative model server. Level 3 could also be regarded as integrated BIM or iBIM potentially employing concurrent engineering processes. Maturity level 3 BIM would essentially mean creating a joined design environment with full collaboration between all disciplines using a single, shared project model which is held in a centralised repository which all members can instantly access and modify the same model.

To achieve higher BIM maturity levels with the integrated environment, it is important to implement BIM on various levels of the educational process and to use it in interdisciplinary courses. Reported case studies of the BIM curriculum design in undergraduate courses are significantly larger than cases of BIM curriculum in graduate courses. Moreover, BIM cases in architecture, architectural engineering and building science courses seem to be fewer than cases of the integration of BIM into civil engineering and construction management courses. (Abdirat and Dossick 2016) Additional elective or regular courses which implement BIM, together with standalone BIM courses at the undergraduate level support students in long-time learning and ability to use BIM skills in different learning environments (Clevenger et al. 2010). BIM is important for filling gaps during the building delivery process by providing a virtual repository that allows easy access to, and sharing of, information and knowledge in real time. It provides a holistic understanding of how choices made on one building system will affect another or more building systems. (Fadeyi 2017) Future BIM pedagogical work should involve more project-based case studies with real-world scenarios application in the pedagogy and industrial professionals providing constructive feedback to students, in turn contributing design proposals to project stakeholders for their appraisal (Jin et al. 2018).

BEM IN SUSTAINABLE DESIGN PROCESS

Building energy performance related to energy efficiency can be greatly influenced by the earlier design of a building, i.e., building orientation and shape, window-to-wall ratios, thermal envelope, zoning, ventilation, shading etc. Energy evaluation and life-cycle performance, HVAC systems and living comfort are usually assessed and implemented after the design phase. However, with the help of new digital tools, the design of energy-efficient buildings can be directly linked to architectural design. BIM has the capability of storing multidisciplinary information within one model and creates the opportunity for sustainability analysis to be performed at initial design stages (Schlueter and Theseeling 2009).

In the traditional energy performance process, an energy specialist manually creates an energy model using an architect's drawings, other specifications, and photos. This method, which largely depends on the specialist's proficiency, has a high probability of data loss, inconsistency, and inaccuracy when simplifying the available building design information for an energy model, and includes numerous possibilities for error and unreliable analysis (Choi et al. 2016). Therefore, a BIM-based energy performance assessment could be a good solution to optimising the interoperability between the designed building and energy performance analysis and achieving valid results.

For building energy performance assessment, BIM model is adapted to building energy model (BIM-to-BEM). For BIM-to-BEM technological approaches, three integration methods can be used: combined, central, and distributed method. The combined method is performed where design and energy simulation (BIM and BEM models) are completed using one tool, which means that one practitioner (acting both as an architect and an engineer) completes the modelling of the building and simulation (Farzaneh et al. 2019). This method is especially useful for interdisciplinary approaches in architecture education, where students do not have to learn various types of software for modelling, energy simulations, and other performance evaluations. Central and distributed methods use different tools for the integration of BIM and BEM. The central method centralises the building information in a shared environment using a coupling medium or data schema as the interoperability gateway, while the distributed method transfers data between tools using middleware (Farzaneh et al. 2019).

BEM model inputs usually include a description of building properties with geometry and construction materials, site information, buildings systems, such as HVAC, water heating and lighting, the control of building systems, their capacities, and component efficiencies. It also takes into account building use with operation and schedules for occupancy, lighting, equipment, and internal temperature settings. These inputs are combined with information about local weather data. The outputs calculated by physics equations are calculations of thermal loads, system response to these loads, and the resulting energy use along with related metrics like living comfort and energy costs. The energy performance of a given building is calculated and predicted using energy simulation tools with certain criteria based on thermodynamic principles and assumptions. The accuracy of an energy performance assessment depends on the accuracy of the input data for a building. If the input data do not accurately represent a real building, the simulation results could be arbitrary and inaccurate. (Choi et al. 2016) This is also why BIM-to-BEM integration can produce results with fewer mistakes and uncertainties.

The application of BIM in energy-related modelling and simulations provide major benefits in building energy management, such as automation of energy modelling, better presentation of energy-related outputs, the capability of storing and organising data on a building, and also enhancing existing libraries by adding new attributes (for example new materials, their properties etc). (Kamel and Memari 2019) The integration of BIM and BEM helps with the optimisation of a building to achieve high indoor environment comfort with reduced energy consumption and lifecycle cost (Jalilzadehazhari and Johansson 2018), integrated energy optimisation of sustainable buildings (Gourlis and Kovacic 2017) or multicomponent energy assessments (Singh and Sadhu 2019).

APPLICATION IN EDUCATIONAL PROCESS

Our faculty offers standalone BIM courses and courses which implement BIM skills in the curriculum. The course of "Sustainable concepts of building design" is a 2nd year course at the master's architecture study programme and the master's civil engineering study programme. The use of BIM technologies as support for integration in a multidisciplinary course has been used for planning and assessment. The presented case is a student workshop integrated in the educational process with an interdisciplinary approach to teaching in collaboration with the local community.

The design process for the course is divided into a few parts:

- the conceptual design with starting points or guidelines, detailed site analysis, research (programme, zoning);
- the preliminary design with the implementation of sustainable design concepts, energy performance simulations, floor plan studies
- the design development with the analysis of environmental impacts, the design of the envelope, and analyses of energy savings and living comfort etc.

There is a significant difference between student projects and real-time projects, as some phases of the conventional design process, such as documentation, consultation and negotiation, and the construction of the project are not implemented in the study process. However, in the early phases of the design process very similar challenges arise. Usually, in the conventional design process architects, engineers and other team members do not work together in the early design phases. That is why the optimisation of the project is often limited to later phases of the process and frequently very hard or impossible to implement. The cooperation between the architecture and engineering fields in the initial design phases and the integration of digital technologies for easier assessment of the proposed solutions are very important aspects included in the course.

In the conceptual design phase, the use of digital technologies is already implemented for drawings, simple visualization, infographics, concept drawings etc., primarily as a drawing tool. However, at the preliminary design and design development phases, digital technologies are used as planning and assessment tools. BIM is implemented as a tool for architecture design and building energy performance simulation tool, which helps in the preliminary design optimisation. Additionally, it functions as a collaboration tool for the exchange of data between students in different fields.

In undergraduate courses, students already get familiar with simple software for the assessment of the energy efficiency of a building. The main objective of that course is to obtain knowledge on energy flows in buildings, acquire a general feeling of energy used in buildings in Slovenia, and to understand the basic principles of energy-efficient design of buildings. Prior knowledge helps students to better understand the principles for modelling a building for energy analysis, and the settings and parameters of the software. Prior knowledge of energy efficiency and living comfort in buildings is very important for better understanding of complex building. Digital models of a building can contain a lot of detailed information about the building and its site. Therefore, it should be understood that computational models are often simplifications of reality for efficient data processing. To make use easier, certain programmes provide default settings that may not be appropriate for every project case in every environment, and some available data, such as weather or other files, may not be completely accurate for a certain project (Sawyer and Weissman 2012).

CASE STUDY

In the case study presented, architectural software ArchiCAD from the Hungarian company Graphisoft was used from BIM. BIM-based BEM was created directly from the geometry modelled in ArchiCAD by integrated energy evaluation tool EcoDesigner STAR. Energy Evaluation tool integrated in the ARCHICAD environment provides an easy-to-use workflow for dynamic building energy calculations for projects of any size.

It enables architects to monitor and control all architectural design parameters that influence building energy performance at all stages of the design process. (Graphisoft.com 2019b) However, the main idea of the course is not only to teach students how to use this exact software tool, but also to understand the BIM process as an integrative design process with a multi-disciplinary collaboration.

Conceptual and Preliminary Design

The first design ideas were sketched on a paper, or digitally drawn on the computer. The detailed site analysis and the analysis of climate conditions were carried out. The research of the programme and the zoning of the area were implemented. The guideline was to implement sustainable concepts from different aspects. The interdisciplinary approach is in that case very useful in the implementation of a holistic view of sustainable development.

The preliminarily designed objects were modelled in the BIM environment by choosing appropriate building components and materials. Selecting appropriate components and materials can be quite problematic in this phase, since students have to acquire a lot of knowledge of architectural elements, structural components, material properties etc. This student workshop was implemented in the last year of the graduate study in which good prior knowledge is expected.

Energy Simulation Modelling

To produce BEM from BIM, additional settings or elements had to be modelled in the software. Firstly, zones had to be defined, which are spatial elements directly enclosed by building elements – walls, slabs, roofs, and windows or other openings. Those zones represent the volume of a room, the part of the building or a functional area with certain properties, and have to be modelled correctly by automatic recognition of borders.

The BIM example from the student workshop and BEM showing zones of the building are shown in Figure 1.

Secondly, the settings for operation profiles were edited. Daily schedules were adjusted to account for certain building zones which are used only partially or have various types of use. The internal temperature in all analysed buildings was set to a minimum of 20°C and a maximum of 25°C. Occupancy count, lighting, and equipment were adjusted to achieve a total of 4 W/m2 of internal heat gains. The climate for Podlehnik was imported in the software. Another possibility to get climate data is to locate the building site with coordinates and download the climate data for the location from Strusoft Climate server. (Graphisoft.com 2019a) The climatic conditions were already analysed in the conceptual design phase and possible strategies to achieve good energy performance of the buildings were discussed on various levels – the urban design, the building design, and the design of building components.

Figure 1. BIM and BEM model of a multifamily residential building.

The surrounding environment was partially modelled or selected from the dropdown settings to account for wind protection and horizontal shading. Certain connections between building elements were additionally assessed to reduce the losses through the building envelope using a thermal bridge simulator. Building systems had to be selected for the energy performance simulation. For this climate both heating and cooling systems were included, as the climate conditions include cold winters and hot summers. With larger glazing sizes on new buildings, the rooms can easily get overheated. For most buildings, especially the ones in public use, mechanical heat recovery ventilation was included. However, for some buildings, also natural ventilation was incorporated particularly to be used in the spring and autumn.

In the final step before energy performance simulation, the zones were added to certain thermal blocks with the same properties (operation profiles and building systems). All the structures were checked, the U-value calculations for the thermal envelope components were calculated, and the infiltration through building elements was adjusted. Regarding openings, the right windows and door types had to be selected, and the properties of frames and glazing elements adjusted. An additional solar analysis of glazed elements results in the percentage of glazed areas exposed to direct sunlight.

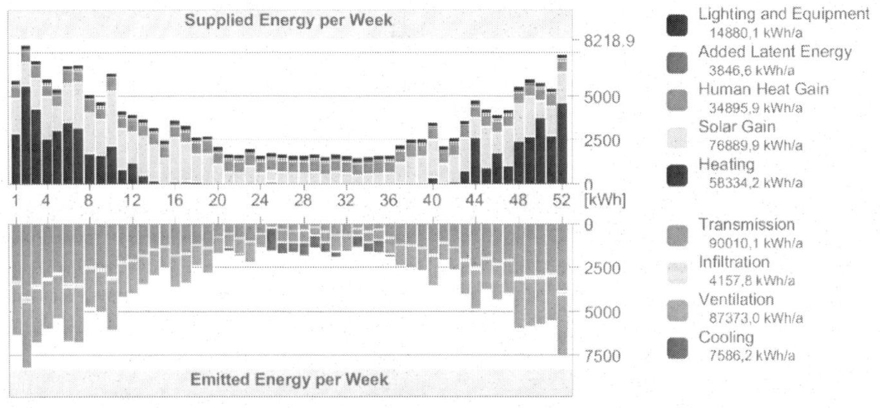

Figure 2. BIM-based BEM energy performance results for multifamily residential building.

After setting up the energy model, the results were analysed, presented, and discussed in groups and with mentors. Energy demands for heating and cooling were calculated and individual energy flows in the buildings were also analysed. To understand heated and unheated period better, an additional assessment was done on the basis of monthly values of energy flows in the spreadsheet results exported from the software. Figure 2 shows the results and analysis for one of the case study buildings.

Design Development

After the preliminary design and energy performance simulations, questions about living comfort in the building were discussed, for example natural illumination of the spaces, possible overheating in some areas, and other aspects. For assessment of the natural light in the building and structural analysis, BIM models were exported to another software. Students of structural engineering proposed changes to the building construction. Therefore, certain adjustments in building components were proposed. Using lighting simulation tools, students predicted and analysed the level of daylight in the space and proposed the enlarging of glazing where necessary.

Since students are already familiar with the energy modelling in this phase, they can easily adjust the settings and the BIM-based model, and try variations in the building design to achieve optimised building performance.

CHALLENGES

Despite the various potential advantages of using BIM and established application strategies, the implementation of BIM in the educational process and in the industry sector remains challenging, and the actual efficiency of using BIM has not been fully identified.

Potential problems associated with the general adoption of BIM in construction projects are initial costs of BIM implementation and concerns about long-term effects and return on investment. Financial and legal risks, the lack of qualified personnel available, and inadequate project experience are just a few of the challenges the AEC industry has been facing in the application of BIM. To gain the benefits of BIM techniques, educators in the AEC disciplines should rethink their teaching methods, and promote and evaluate professional competencies of future AEC professionals. (Ghaffarianhoseini et al. 2017) Certain challenges in the application of BIM-based energy performance simulation in the education process were addressed.

Operating with BIM requires thorough prior knowledge of construction details, processes, and building systems, and demands expertise for the use of advanced digital tools. In interdisciplinary courses or workshops, all students do not have the same prior knowledge of sustainability, energy efficiency, and living comfort. During lectures, principles of sustainable design are explained and discussed again. However, in the use of the software for energy performance simulations, many settings are not understandable to everyone. It is a lot easier if students have already done energy calculations in another software or at least have good prior knowledge of the energy performance of buildings. Uncertainty in understanding can result in inaccurate settings in the software and further false interpretations of the energy performance results.

The second challenge is the difference in prior knowledge of BIM. Despite the emphasis on BIM implementation throughout the study process and other courses, some students still use traditional CAD planning environments, because they feel more accustomed to them or they do not take the time to properly study BIM. In the last few years, BIM has been increasingly implemented in various courses in the first years of the study programmes of architecture and civil engineering. Therefore, the percentage of students using BIM as a primary planning programme has risen and this issue has already improved.

Working in interdisciplinary teams should not only be encouraged between students of the AEC professions, but should also be extended to collaborations with other faculties to achieve even better skill sets and experience using advanced digital tools. In practice, AEC professionals still face the problem that other professions like electrical engineers, mechanical engineers or other consultants do not use the same software tools or do not even use BIM tools at all. As a result, the link between professions and the level of maturity in BIM use is still very low. Educational institutions promote and generate the transformation of work in practice. Working in interdisciplinary groups helps students to learn from each other, better understand the responsibilities and obligations of other fields, and handle the tasks effectively. It is also easier for multidisciplinary teams to learn new software tools and use them accordingly in the sustainable design process.

CONCLUSION

This paper presents the integration of digital tools in the building design process. Digital technologies have been changing the architectural design process and the ways buildings are designed, while providing opportunities for the collaboration of multidisciplinary team members.

It is stated that BIM, as an emerging technology and as a process, promotes changes in the traditional design process by encouraging iterative design and multidisciplinary interaction. The BIM models can be linked to other software tools, saving time for additional research, model evaluations, and optimisations. The BIM-based building energy modelling enables energy performance simulations in the early design phase. It helps to achieve a more cost-effective and time-efficient design of energy-efficient buildings with easier design optimization and improved results review.

The implementation of BIM and BIM-based energy modelling in the educational process can provide experience with interdisciplinary work and advanced digital tools. The presented case study of an interdisciplinary student workshop in the second year of the master's study programme at the University of Maribor is an example of the integration of BIM-based energy modelling in the sustainable design process. We highlighted some of the challenges of implementing BIM technology in the education process and proposals for achieving higher levels of BIM knowledge and maturity. Good knowledge of construction processes and details, a clear understanding of the principles of sustainable design, and the prior use of BIM are just some of the practices required. With continuous modifications and improvements over the years, the course with the project-based learning approach can be considered a successful learning experience for the teaching of BIM-based energy simulations in an interdisciplinary environment.

The role of architects in the sustainable design process has been visibly changing and evolving from knowing how to communicate design ideas to a wider set of skills in management and construction. It is still evident worldwide that architects are usually educated separately from engineers, engineers separately from structural engineers, structural engineers separately from economists etc. To be able to implement BIM techniques and multidisciplinary collaborations in a clear way, it is necessary to insist on education in a multidisciplinary environment by linking different professions, working on real-world projects, and introducing decision-making in a dynamic environment.

To sum up, it is very important to integrate BIM and multidisciplinary work into the educational process, as it emphasises the knowledge of different disciplines, and often improves constructive dialogue and interactions. There are, of course, challenges with the implementation of BIM and other digital technologies in the educational process and in practice. However, with a positive attitude towards their integration, we can effectively address changes in the sustainable design process.

References

A report for the Government Construction Client Group Building Information Modelling (BIM) *Working Party Strategy Paper*. (2011). (online) Available at: https://www.cdbb.cam.ac.uk/system/files/documents/BISBIMstrategyReport.pdf (Accessed 30 Aug. 2019).

Abdirat, Hamid and Dossick, Carrie S. (2016). BIM curriculum design in architecture, engineering and construction education: A systematic review. *Journal of Information Technology in Construction*. 21: 250-271.

Al-Qawasmi, Jamal and Hadjri Karim. (2007). Architecture in the digital age: The effect of digital media on the design, production and evaluation of the built environment. *Open house international*. 32(2): 4-6.

Angulo, Antonieta and Vásquez de Velasco, Guillermo. (2007). Digitally integrated practices: a new paradigm in the teaching of digital media in architecture. *Arquiteturavista* 3(2): 1-14.

Azhar, Salman; Carlton, Wade A., Olsen, Darren, Ahmad, Irtishad. (2011). Building information modelling for sustainable design and LEED® rating analysis. *Automation in Construction* 20: 217-224. doi:10.1016/j.autcon.2010.09.019.

Azhar, Salman; Khalfan, Malik and Maqsood Tayyab. (2012). Building information modelling (BIM): Now and beyond. *Australasian Journal of Construction Economics and Building* 12(4): 15-28.

Bagheri, Ali and Hjorth, Peder. (2007). Planning for Sustainable Development: a Paradigm Shift towards a Process-Based Approach. *Sustainable Development* 15:83-96.

Braungart, Michael and McDonough, William. (2013). *The Upcycle: Beyond Sustainability--Designing for Abundance.* New York: North Point Press.

Choi, Jungsik, Shin, Jihye, Kim, Michan, Kim, Inhan. (2016). Development of openBIM-based energy analysis software to improve the interoperability of energy performance assessment. *Automation in Construction* 72: 52-64. doi:10.1016/j.autcon.2016.07.004.

Clevenger, Caroline M., Ozbek, Mehmet E., Glick, Scott, Porter, Dale. (2010). *Integrating BIM into Construction Management Education. EcoBuild 2010 BIM Academic Forum.* Washington DC.

Dewberry, Emma and Fletcher, Kate. (2001). Demi: linking design with sustainability. In *the 7th European Roundtable on Cleaner Production IIIEE*, May 2-4. Lund, Sweden.

Fadeyi, Moshood Olawale. (2017). The role of building information modelling (BIM) in delivering the sustainable building value. *International Journal of Sustainable Built Environment* 6: 711-722.

Farzaneh, Aida, Monfet, Danielle and Forgues, Daniel. (2019). Review of using Building Information Modeling for building energy modelling during the design process. *Journal of Building Engineering* 23: 127-135.

Gao, H., Koch, C., Wu, Y. 2019. Building information modelling based building energy modelling: A review. *Applied Energy* 238: 320-343. doi:10.1016/j.apenergy.2019.01.032.

Ghaffarianhoseini, A., Tookey, J., Ghaffarianhoseini, A., Naismith, N., Azhar, S., Efimova, O. and Raahemifar, K. (2017). Building Information Modelling (BIM) uptake: Clear benefits, understanding its implementation, risks and challenges. *Renewable and Sustainable Energy Reviews* 75: 1046–1053. doi:10.1016/j.rser.2016.11.083.

Gourlis, Georgios and Kovacic, Iva. (2017). Building Information Modelling for analysis of energy efficient industrial buildings – A case study. *Renewable and Sustainable Energy Reviews*. 68: 953-963. doi:10.1016/j.rser.2016.02.009.

Graphisoft.com. (2019a). *Climate Data (Energy Evaluation) | User Guide Page | GRAPHISOFT Help Center*. (Online) Available at: https://helpcenter.graphisoft.com/user-guide/77230/ (Accessed 2 Sep. 2019).

Graphisoft.com. (2019b). *Energy Evaluation | User Guide Page | GRAPHISOFT Help Center*. (Online) Available at: https://helpcenter.graphisoft.com/user-guide/65686/ (Accessed 30 Aug. 2019).

Gross, Mark D. and Do, Ellen. (1999). Integrating Digital Media in Design Studio: Six Paradigms. Proc. ACSA (American Collegiate Schools of Architecture) *National Conference '99*. Minneapolis, Minn.

Hasanpour Loumer, Saeid. (2015). Digital Architecture and Intelligent Buildings: A Suitable Approach to Proper Implementation of Sustainable Development Components in the Third Millennium. *Current World Environment* 10:74-81.

Jalilzadehazhari, Elaheh and Johansson, Peter. (2018). Integrating BIM, Optimization and a Multi-criteria Decision-Making in Building Design Process. Advances in Informatics and Computing in Civil and Construction Engineering. *Proceedings of the 35th CIB W78 2018 Conference.* Springer Nature Switzerland.

Jin, Ruoyu; Yang, Tong; Piroozfar, Poorang; Kang, Byung-Gyoo; Wanatowski, Dariusz; Hancock, Craig Matthew and Tang, Llewellyn. (2018). Project-based Pedagogy in Interdisciplinary Building Design Adopting BIM. *Engineering, Construction and Architectural Management*. 25(10): 1376-1397. doi: 10.1108/ECAM-07-2017-0119.

Kamardeen, Imriyas. (2010). 8D BIM modeling tool for accident prevention through design. Paper presented at the *26th Annual ARCOM Conference*. 2010 Sep 6-8; Leeds: Association of Researchers in Construction Management.

Kamel, Ehsan and Memari, Ali M. (2019). Review of BIM's application in energy simulation: Tools, issues, and solutions. *Automation in Construction* 97: 164-180. doi:10.1016/j.autcon.2018.11.008.

Kara, Levent. (2014). A critical look at the digital technologies in architectural education: when, where and how? *Procedia – Social and Behavioral Sciences* 176: 526-530. doi:10.1016/j.sbspro.2015.01.506.

Kushwaha, Vinay. (2016). Contribution of Building Information Modeling (BIM) To Solve Problems In Architecture, Engineering and Construction (AEC) Industry and Addressing Barriers to Implementation of BIM. *International Research Journal of Engineering and Technology (IRJET)* 3(1): 100-105.

Latiffi, Aryani Ahmad; Brahim, Juliana and Fathi, Mohamad Syazli. (2014). The Development of Building Information Modeling (BIM) Definition. *Applied Mechanics and Materials* 567: 625-630. doi:10.4028/www.scientific.net/AMM.567.625.

Larsson, Nils. (2009). The Integrated Design Process; History and Analysis. International Initiative for a Sustainable Built Environment (iiSBE). http://www.iisbe.org/node/88.

Leerberg, Malene, Riisberg, Vibeke and Boutrup, Joy. (2010). Design Responsibility and Sustainable Design as Reflective Practice: An Educational Challenge. *Sustainable Development* 18: 306-317. doi:10.1002/sd.481.

McCrory, R. J. (1966). The design method in practice. Published in: Gregory, Sydney A., *The design method*, Springer: London.

Sakikhales, Mohammad H. and Stravoravdis, Spyros. (2015). Using BIM to Facilitate Iterative Design. *Building Information Modelling (BIM) in Design, Construction and Operations* 149: 9-19.

Sawyer, Azdeh O. and Weissman, Daniel A. (2012). Design with Climate: The role of Digital Tools in Computational Analysis of Site-specific Architecture. Published in: Hanif, Kara and Georgoulias, Andreas. 2012. *Interdisciplinary Design*. New York: Actar.

Schlueter, Arno and Theseeling, Frank. (2009). Building Information Model Based Energy/Exergy Performance Assessment in Early Design Stages. *Automation in Construction* 18: 153-163.

Singh, Premjeet and Sadhu, Ayan. (2019). Multicomponent energy assessment of buildings using building information modelling. *Sustainable Cities and Society* 49: 1-12.

World Commission on Environment and Development (WCED). (1987). *Our Common Future*. Oxford: Oxford University Press.

In: An Interdisciplinary Approach … ISBN: 978-1-53617-302-4
Editors: V. S. Klemenčič et al.

Chapter 5

STRUCTURAL DESIGN OF SUSTAINABLE TIMBER BUILDINGS

Miroslav Premrov[*], *PhD*
University of Maribor, Faculty of Civil Engineering, Transportation Engineering and Architecture, Maribor, Slovenia

ABSTRACT

The first part of this chapter presents a set of main aspects of timber building by briefly describing the main types of traditional structural systems with a primary focus on massive-panel (CLT) and timber-framed construction. In the second part, sustainable impact on the design of contemporary prefabricated timber buildings is presented by integrating two quite different building materials; timber as a raw natural material with an excellent CO_2 footprint, and glass as a transparent material with many possibilities to improve energy efficiency of timber buildings. However, attractive timber-glass buildings designed in this way can cause many structural problems, especially in the sense of racking resistance on windy or seismic areas. In this sense and to avoid this, special prefabricated timber-glass wall elements were developed through many previous studies

[*] Corresponding Author's Email: miroslav.premrov@um.si.

and international research projects briefly presented at the end of the chapter, which can open many other opportunities in further development of multi-story prefabricated timber buildings with enlarged glazing areas.

Keywords: timber, structural systems, glass, timber-glass elements

INTRODUCTION

As a natural raw material timber shows indisputable environmental excellence and is certainly one of the best choices for sustainable construction. Moreover, timber has good mechanical properties and ensures a comfortable indoor climate in addition to playing an important role in the reduction of CO_2 emissions. Trees absorb CO_2 while growing (estimated CO_2 absorption of conifers is approximately 900 kg per 1m^3 with that of deciduous trees being 1,000 kg per 1m^3), which makes timber carbon neutral. Building made of an adequate mass of timber can thus have even a negative carbon footprint.

Table 1 shows grey energy consumption for production of the element for most frequently used building materials and their end-product elements. As the density of the materials varies, the values of energy consumption per kg and m^3 are given separately (Žegarac Leskovar and Premrov 2013).

Table 1. Grey energy consumption for selected building materials

Building material	Grey energy [MJ/kg]	Grey energy [MJ/m^3]
Timber	0.3–1.6	165–638
Wood-based board products	8.0–11.9	5,720 –5,694
Brick	2.5–7.2	5,310–14,885
Cement	5.2–7.8	12,005–12,594
Glass	15.9	40,039
Steel	31.3–74.8	245,757–613,535
Aluminium	191–227	517,185–611,224
Aluminium – recycled	8.1–42.9	5,792–24,397
Insulation l' polystyrene	117	1,401

It is evident from the presented data that grey energy consumption in producing 1 kg of the timber element is the lowest of all, having a value nearly five times lower than that of brick, six times lower than in the case of cement, and approximately 50 times lower in comparison to steel.

In comparison to other types of buildings, the energy-efficient properties of prefabricated timber buildings are excellent not only because well-insulated buildings use less energy for heating, which is environmentally friendly, but also due to a comfortable indoor climate of timber-frame houses (Gold and Rubik 2009). Considering the growing importance of energy-efficient building methods, timber construction will play an increasingly important role in the future. The use of timber in construction is gaining ever more support, especially in regions with vast forest resources, since it reduces energy demand for transport if the building material is available from the local area. With respect to all the given facts, timber as a material for load-bearing construction poses a future challenge for residential and public buildings.

Besides the listed ecological benefits, there are currently further additional strong arguments in favour of building timber structures. Brand new and improved features introduced in the early 1980s brought about a significant expansion in timber buildings all over the world. In addition to the listed ecological benefits, there are currently additional strong arguments in favour of building timber structures, as seen from the structural and technological points of view. Brand new and improved features introduced in the early 1980s brought about a major expansion of prefabricated timber buildings all over the world.

The most important changes introduced in prefabricated timber construction during the last 50 years are listed below:

- Transition from on-site construction to prefabrication in factory under controlled climatic conditions.
- Transition from elementary measures to modular building.
- Improvement of the carbon footprint of prefabricated elements by using eco-friendly thermal insulation.

- Development from the single-panel wall system to the macro-panel wall prefabricated system, which essentially increased the time of construction erection.

New timber products in the last decades (cross-laminated timber, for instance, at the beginning of the 1990s) changed the form and especially the height of timber buildings, and timber structures became competitive with other structures built with classical building materials (especially with brick and concrete structures). Today, timber buildings with a height up to 20 stories can be built in the cross-laminated structural system. Furthermore, combining timber structural elements with other building materials (for instance, with glass, brick, concrete or steel) can open new perspectives on attractive architectural forms of such hybrid timber buildings. Especially a combination with load-bearing glass elements will therefore be presented at the end of this chapter in the sense how to build an energy efficient timber building with enlarged glazing areas by taking advantage of the passive strategy in heating time and consequently in the same time to satisfy all structural stability requirements. Such interacted architectural–structural approach is very important to define a timber building in a view of an optimal indoor living comfort.

Schematic presentation of the structural systems of timber buildings is presented in Figure 1. The classification is made on their load-bearing function. However, prefabricated timber construction systems differ from each other in the appearance of the structure, and in the approach to planning and designing a particular system. As presented in (Kolb 2008) and (Žegarac Leskovar and Premrov 2013), structural systems of timber buildings can be classified into six main systems:

- Log construction (Figure 2a),
- Solid timber construction (Figure 2b),
- Timber-frame construction (Figure 2c),
- Frame construction (Figure 2d),
- Balloon- and platform- frame construction (Figure 2e),
- Frame-panel construction (Figure 2f).

Log construction and solid timber construction can also be classified as massive structural systems, since all load bearing elements consist of massive solid elements. Other construction systems consist of timber-frame bearing elements and are therefore classified as lightweight structural systems. According to the load bearing function, they can be subdivided into classical linear skeletal systems (Figures 2c and 2d), in which all the loads are transmitted via linear bearing elements and planar frame systems (Figures 2e and 2f), in which sheathing boards take over horizontal loads and structural wall elements can be treated as two-dimensional planar components.

It is important to point out that only solid timber construction (Figure 2b) and frame-panel construction (Figure 2f) are prefabricated, other structural systems are classical and usually built on-site. Therefore, most of our attention will be further focused on these two structural systems to analyse most efforsts of sustainable prefabricated timber building.

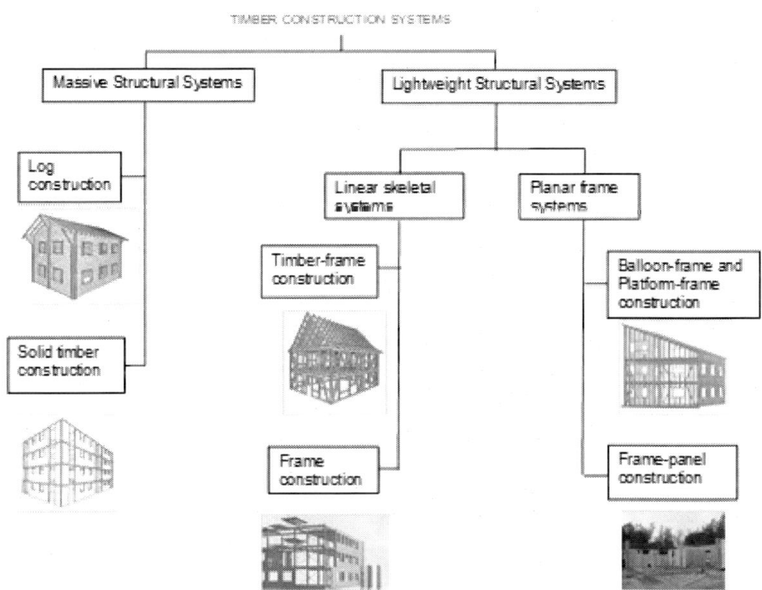

Source: Žegarac Leskovar and Premrov 2013.

Figure 1. Classification of timber structural systems according to their load-bearing function.

a) Log construction
b) Solid timber construction
c) Timber-frame construction
d) Frame construction
e) Balloon-frame construction
f) Frame-panel construction

Figure 2. Main structural systems in timber building.

SOLID TIMBER CONSTRUCTION

A solid timber structural system is a prefabricated massive panel timber construction where the load-bearing wall and floor elements are produced as cross-laminated planar structural elements. The main benefit of this prefabricated cross-laminated structural system is in the perpendicular orientation of timber boards (Figure 3a) to avoid anisotropy of wood as a raw natural material (Žegarac Leskovar and Premrov 2013). The boards with the maximal width of 240 mm are posted in layers perpendicular to each other (Figure 3a) and then glued with the polyurethane adhesive (Figure 3b). Therefore the name "cross-laminated timber" (CLT) or "Kreuzlagenholz – KLH" or "X-lam" is usually used. The number of layers is always even. The basic material for the production of CLT-elements is sawmill-boards, whose quality is best if they are cut from the outer zones of the log. Such side-boards, which are not considered as particularly profit-making items by the millers, generally have excellent mechanical properties relating to stiffness and strength, with their thickness ranging from 10 to 45 and up to 100 mm (depending on the producer). The width to thickness ratio should be defined as *b: d = 4:1*. Timber species currently processed belong to conifers (e.g., fir, spruce fir, pine), which does not exclude deciduous trees (e.g., ash, beech) from being used in the future, (Augustin 2008).

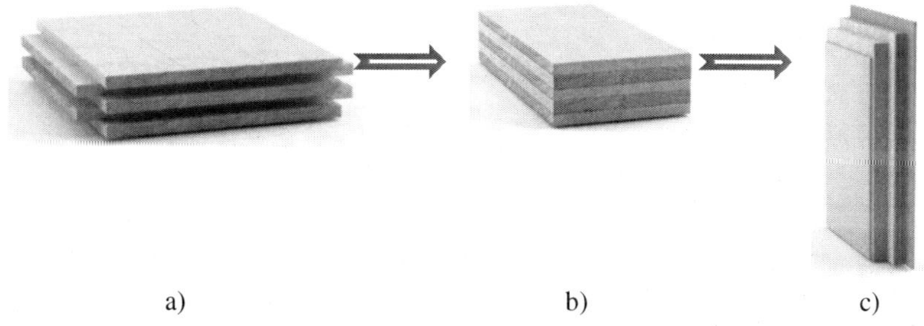

Figure 3. Composition of cross-laminated timber structural elements.

The elements can be produced as the structural resisting wall or floor elements. In case of an external wall element, the structural part of the element is completed with an additional layer of thermal insulation to decrease the thermal transmittance (U-value) of the envelope element (Figure 3c). Insulation in CLT wall elements is placed on the external sides of the bearing timber construction. This is one of the most significant disadvantages of CLT wall elements in comparison to frame-panel wall elements. Gypsum fibreboards are added on the external sides of the elements to improve fire resistance of wall elements.

It is of the utmost importance to provide stiff anchoring of all resisting CLT wall elements to bottom floor elements (Figure 4a) as well to the concrete foundation plate (Figure 4b). In this case, only the horizontal load actions (wind or seismic loads) can be transferred to the load-bearing elements of the lower story (case a) and finally to the foundation (case b). However, at the same time, these steel connectors can dissipate the energy and therefore ensure a highly ductile design of such type of timber structures with a relatively high range of seismic resistance.

The size and form of CLT elements are defined by restrictions concerning production, transport, and assembly. The existing standard dimensions of planar and single-curved elements are set at a length, width and thickness of 16.5 m, 3.0 m and up to 0.5 m, respectively. Longer elements (of up to 30 m) can be assembled by means of general finger-joints. The thickness of lamellas in curved CLT elements has to be adjusted to the curvature (Augustin 2008).

a) b)

Figure 4. (a) Steel shear connector of the wall element to the bottom floor, (b) Steel shear connector of the 1st story wall element to the foundation plate.

Today most of the highest "non-hybrid" timber buildings are built mostly in CLT structural system. Nine-story prefabricated timber building built in CLT structural platform system only as a residential tower block in Hackney, East London. On its completion in 2008 "Stadthaus" was the tallest timber residential building in the world. A building of this form would normally be produced as a reinforced concrete frame, an energy intensive process producing upwards of 125.000 kg of carbon. In contrast Stadthaus actually stores 185.000 kg of carbon within the timber structure (Waugh et al. 2009). The 10-story CLT building, presently regarded as the tallest timber apartment building in the world, was built in Melbourne. It was demontrated on this project that timber high-rise buildings are a good answer to sustainable building in urban areas. However, it was also found out that the cost of building is somewhat higher than in steel or concrete structural design, but erection time is shorter. Additionally, it is important to stress that the carbon footprint is favorable.

Figure 5. Eight-story CLT and timber-framed buildings Limnologen (Vaxjo).

The structural design of tallest timber buildings was afterward turned to hybrid structural systems in timber only. An interesting approach using a combination of CLT and timber-framed load-bearing wall elements in multi-story timber buildings was first used in Limnologen (Vaxjo, Sweden). It included four eight-story blocks of flats built in 2008 (Figure 5). The buildings consist of seven timber stories on a concrete foundation and of the concrete first floor. The load-bearing structure of external walls consists of CLT elements, while for the wall elements that separate flats timber-framed wall elements were used. In some parts of the buildings, glulam columns and beams supplemented the load-bearing structural system to reduce vertical deformations.

Next such realised example is the 14-story appartment building in Bergen, Norway, finished in 2015. The total height of the building is 52.8 m. The building stands on the top of a concrete garage. It is interesting that the timber structure is designed as a combination of the CLT core and the frame external structure. Totally 550 m3 glulam linear elements are used for the frame structure and 385 m^3 for the CLT core. The glulams are of the typical dimensions of columns 405 mm x 650 mm and 495 mm x 495 mm. Typical all diagonal elements are of square dimensions of 405 mm x 405 mm. It is important to point out that the glulam frame carrie all vertical load, while the horizontal load is carried with a combination of CLT core and the glulam frame. However, the building is not designed for seismic loads. It's so tall that the wind load prevails, which means that seismic design can be omitted according to Norwegian code (Abrahamsen 2014).

Frame-Panel Timber Construction

The frame-panel system originates from the Scandinavian-American construction methods, i.e., balloon-frame and platform-frame construction types, whose assembly takes place on-site. The advantages of the frame-panel construction system over the above-mentioned traditional timber-frame construction systems were first noticed at the beginning of the 1980s and made a significant contribution to the development of such timber construction. However, because of a relatively low racking resistance and especially stiffness comparing with CLT wall elements there are still some limitations in the height and number of stories of timber-framed buildings. The six-story building Hotel Terme Zreče was constructed in year 2000 on Pohorje Mountain in Zreče, Figure 6a. It is important to set out that the lowest two stories (the first is underground) are constructed in massive concrete wall system ande the upper four in timber-framed. The resisting wall elements in the lowest story are reinforced with thin steel diagonal strips which have to be fixed to the timber frame to transfer horizontal loads to the concrete slab. In this case only a part of the horizontal force is shifted from the sheathing boards via the tensile steel diagonal to the timber frame, after the appearance of the first crack in the tensile zone of the fibre-plaster boards.

The benefits to be pointed out lie in factory prefabrication assuring the so called ideal weather condition in addition to constant supervision of construction works and built-in materials (Figure 7). Another asset is the subsequently faster assembly process as the ready-made elements are crane-lifted to be erected onto the foundation platform, adjusted, and screw-fastened.

a) b)

Figure 6. (a) Hotel Dobrava Zreče, (b) Connection of the steel strips to the concrete slab.

Figure 7. Production of prefabricated wall elements under controlled conditions in a factory.

Besides having good constructional characteristics, the frame-panel construction system has a series of additional advantages. The indoor climate of timber buildings is an advantage of major importance since the built-in materials are natural, people-friendly, and demand very little energy for their supply. The panel construction system's priority over massive masonry construction is seen in thinner wall elements, which can result in an increase in the usable floor area by 10 per cent, while the total surface area of a building is the same (Žegarac Leskovar and Premrov 2013). Furthermore, the transition from the single-panel construction system (Figure 8a) to the macro-panel construction system (Figure 8b) in the mid-1990s means even shorter assembly timeand higher stiffness of the entire structure due to fewer joints. In the macro-panel system wall elements with a total length of up to 12.5 metres are now entirely produced under fully controlled conditions in a factory. Typical single-panel dimensions have a constant typical width of b = *1,200 mm* or *1,250 mm,* which is the standard dimension of sheathing boards and a height of h = *from 2,500* to *3,100 mm*. Prefabricated timber-frame walls functioning as the main vertical bearing capacity elements for vertical as well as horizontal load impact are composed of a timber frame and sheets of board-material fixed by mechanical fasteners to both sides of the timber frame. There are many types of panel sheet products available, which may have a certain level of structural capacity, such as wood-based materials (plywood, oriented strand board, hardboard, particleboard, etc., or fibre-plaster boards). Experimental and numerical analysis of the influence of the sheathing board type on load-bearing capacity and racking stiffness of the single-panel wall elements are presented in many studies (Premrov and Kuhta 2009, Premrov and Kuhta 2010, Premrov and Dobrila 2012b).

Wall elements are usually prefabricated in standard modular dimensions, with commonly used 625 mm spans between timber studs (Figure 9). However, to satisfy architectural design requirements for special forms of buildings, exceptions can be made in, for example, window and door openings, and fixed glazing areas. When such exceptions take place, the wall elements next to the opening or those at the end of the whole assembly are usually cut to get the required dimension.

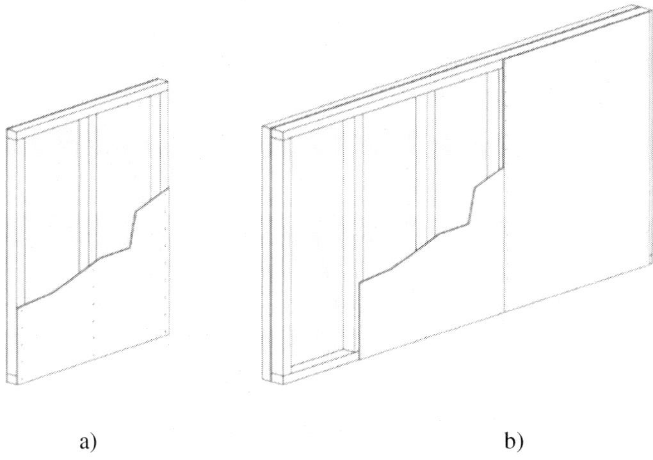

Figure 8. Single-panel (a) and Macro-panel (b) timber-framed wall system.

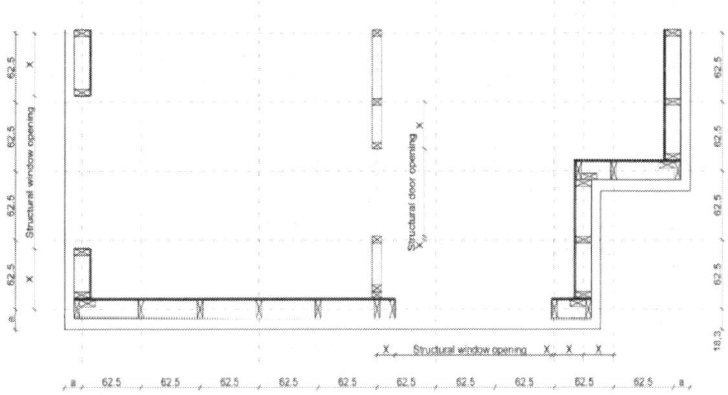

Figure 9. Modular dimensions in prefabricated frame-panel wall elements, dimensions in cm.

From the statical point of view, each wall assembly at individual levels consists of separate wall segments acting as individual cantilevers, where every segment is determined with the width b of a sheathing board (usually 1,200 or 1,250 mm). The lateral forces acting at the top of the element are considered to be equally distributed among segments. The horizontal force acting on a single wall element can be calculated as $F_H = F_{H,tot} / n$. Since, according to Eurocode 5 EN 1995-1-1 (2005), only the segments with the full wall height having no window and door openings are usually taken into account in the calculation, n stands for the number of single-panel wall elements without any openings or fixed glazing. Stress distribution with a horizontal load action in a single-panel wall element is schematically presented in Figure 10. It shows that basically three possible criteria exist to determine the wall racking resistance:

The shear stresses in the shear connecting area between a sheathing board and timber frame elements reach the yielding point of fasteners (Eurocode 5, Method A) applicable mostly to wood-based sheathing boards.

- The tensile stress in a sheathing board reaches the tensile strength of boards (composite wall model) applicable mostly to gypsum fibre-boards.
- The tensile stress in timber frame elements reaches the tensile timber strength.

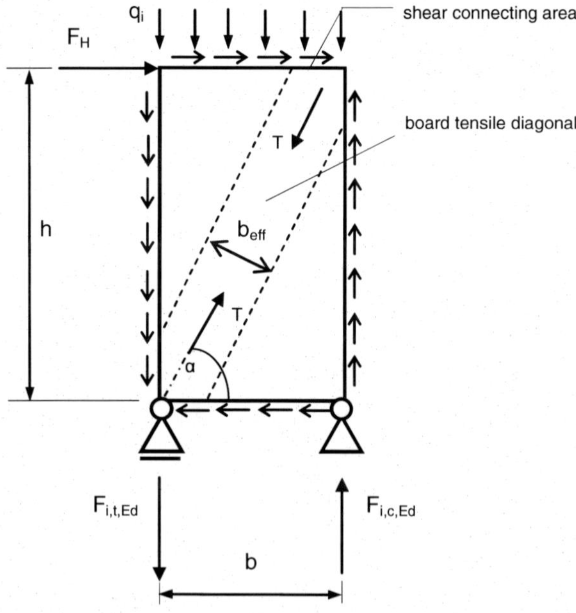

Source: Žegarac Leskovar and Premrov 2013.

Figure 10. Stress distribution in a single-panel frame-panel wall element.

Static Modelling of Resisting Frame-Panel Wall Elements

For static modelling of resisting frame-panel wall elements under a horizontal point load (wind load, seismic actions), a simple static model with one fictive diagonal supported by two rigid supports was developed (Vogrinec and Premrov 2013). The method is based on the equality of the horizontal displacements u_H of the actual timber-framed wall and the model with a fictive diagonal, when the same horizontal force F_H acts on both systems, as schematically shown in Figure 11. The main assumption taken into account is that the horizontal stiffness of the actual timber-framed wall element (Figure 11a) and the model with a fictive diagonal (Figure 11b) is the same. Additionally, it was assumed that the axial stiffness of the surrounding frame is significant enough to eliminate the impact of the frame, and only the flexibility of the fictive diagonal is considered.

Based on these assumptions, the fictive diagonal cross-section $A_{d,fic}$ is developed in final form of:

$$A_{d,fic} = \frac{K_w \cdot L_d}{E_d \cdot \cos^2 \alpha} = \frac{K_w \cdot \left(B^2 + H^2\right)^{\frac{3}{2}}}{E_d \cdot B^2} \quad \text{with} \quad \cos\alpha = \frac{B}{L_d} \tag{1}$$

where K_w is the total horizontal stiffness of the timber-framed wall calculated analytically with the composite cross-section cantilever model (Premrov and Dobrila 2012a), E_d is the modulus of elasticity of the diagonal, B is the width, and H is the height of the fictive diagonal element, and L_d is the length of the diagonal, as schematically presented in Figure 11b. If the circular cross-section of the diagonal element is chosen, its diameter d_{fic} is finally defined as:

$$d_{fic} = 2\sqrt{\frac{A_{d,fic}}{\pi}} \tag{2}$$

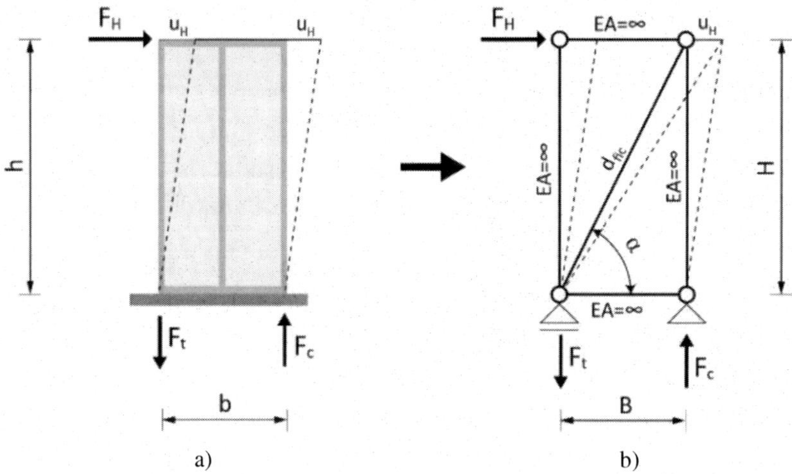

Figure 11. Static model with a fictive diagonal. (a) Actual timber-framed wall element, (b) Static model with a fictive diagonal.

The described model is suitable for the static and dynamic analysis of timber-frame structures, as the three-dimensional model of the building structure can be generated. Forces to each timber-framed wall are distributed in the stiffness ratio, and as the spatial model is being used, irregularities of the structure are also taken into account (and, consequently, torsional effects, etc.). Each timber-framed wall is simulated with the fictive diagonal model. The horizontal force on each wall F_H is calculated as:

$$F_H = F_d \cdot \cos\alpha \tag{3}$$

where F_d is the axial force on a fictive diagonal and $\cos\alpha$ is the inclination of the diagonal as schematically shown in Figure 11b.

The described simple calculation model can be a useful analysis tool for the static and dynamic/seismic analysis of multi-story timber frame structures. It can be used to determine the lateral force distribution on a timber-framed wall for practice engineers. The model provides a relatively quick modelling, and simple and affordable static and dynamic programmes can be used. This simple method is suitable for 3D-modelling (Figure 12a) of timber-framed structures, in which each wall is modelled with the help of the presented model with the calculated fictive diagonal cross-section. The lateral force is distributed to each wall on the basis of the horizontal stiffness of the wall. The whole system behaves as a composition of separate, independent cantilever beams, as schematically shown in Figure 11b, according to the segmented shear wall methods for calculating the force on each wall segment. As the spatial model 3D-model can be used, also irregular structures and the torsion impact can be taken into account. The model also takes into account additional displacements due to the tensile and compressive support, and can optionally consider the impact of openings and cracks in fibre-plaster boards if it is necessary or the designers wish to take them into account. All these characteristics of wall elements are adjusted by optional varying/reducing of the diameter of a fictive diagonal.

If the structure satisfies the prescribed conditions for a regular building in Eurocode 8 (2005), the building can be treated as separated two-dimensional wall assemblies where the horizontal stiffness of each wall element is simulated with a fictive diagonal diameter (Figure 12b).

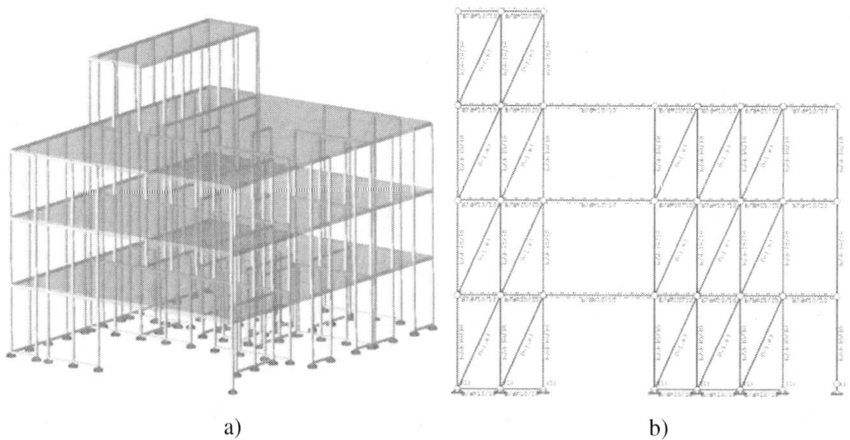

Figure 12. (a) Three-dimensional statical model with a fictive diagonal, (b) Two-dimensional simplification of the 3D-model.

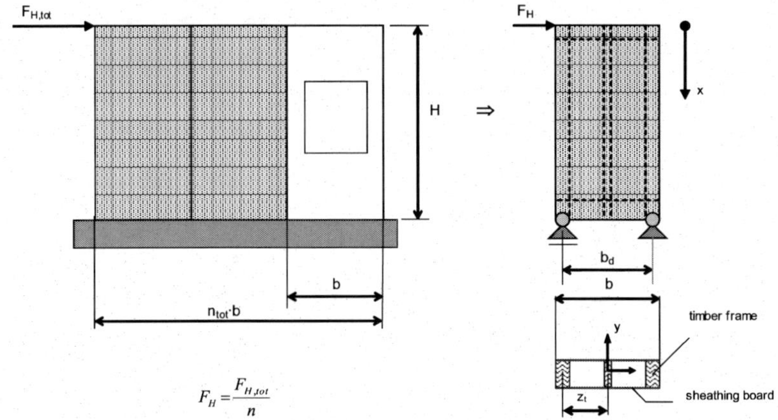

Source: Žegarac Leskovar and Premrov 2013.

Figure 13. Horizontal force distribution in the wall assembly to the single-panel wall element.

The horizontal force acting on one separate timber-framed wall element (F_H) is calculated according to the scheme shown in Figure 13. Only wall elements without any openings or fixing glazing can be taken into account, or optionally the contribution of openings or glazing to the total racking stiffness of the whole two-dimensional wall. However, analysing this relatively new structural approach of so-called timber-glass resisting wall elements is not part of this subchapter and will be briefly presented in the subchapter Timber-Glass Buildings. More details about an in-depth experimental analysis and consequent mathematical modelling supported with a parametric numerical analysis about this may be found in (Premrov et al. 2014, Ber et al. 2018, Štrukelj et al. 2015, Ber et al. 2016).

Prefabricated Timber-Framed Floor Elements

Prefabricated timber-framed floor elements with a typical producing width of 1,000 mm to 1,300 mm usually consist of three to five timber beams with a height varying between 200 mm and 280 mm. Usually, they are covered with a wood-based sheathing board (plywood, particleboard, OSB, etc.) at the top to ensure bending stiffness at the top layer between the beams, and with a fibre-plaster board (FPB) with typical thickness of 15 mm at the bottom to ensure fire resistance (Figure 14a,b). The lower FPB can be also doubled, usually 15 mm + 10 mm.

In the static design of floor elements, the continuous beam system with a vertical dead and live load is usually used (Figure 15). For floors in blocks of flats, the value of $p = 2\ kN/m^2$ is prescribed for live load in Eurocode 1 (2002), while the value of dead load (g) usually varies between 1.2 kN/m^2 and 1.8 kN/m^2. The position of walls can usually be approximated with vertical rigid supports.

For residential buildings with the live load of 2 kN/m^2, maximal spans for such prefabricated timber-framed floor elements are usually up to six metres. In some cases, a span of up to seven metres can be achieved by using five beams with the height of 280 mm. For longer spans, prefabricated CLT floor elements should be used instead of timber-framed elements.

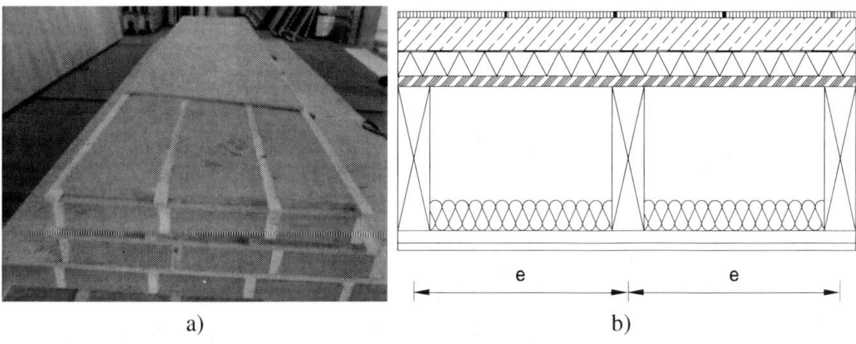

Figure 14. Photo (a) and cross-section (b) of a prefabricated timber-framed floor element.

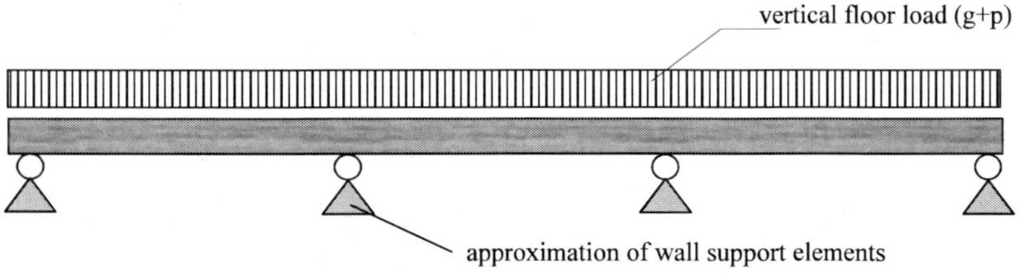

Figure 15. Simplified static design of prefabricated floor elements.

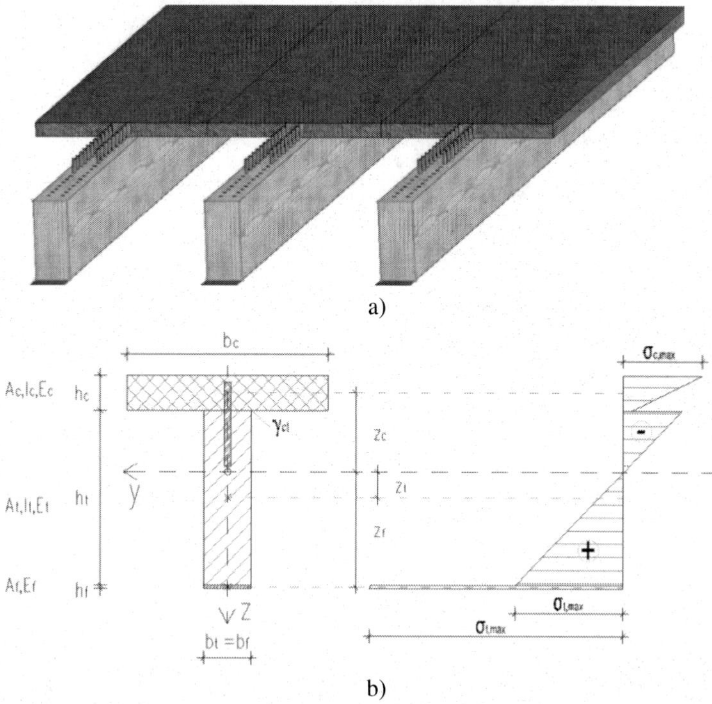

Source: Premrov and Dobrila 2012b.

Figure 16. a.) Composite timber-concrete floor element, b.) Stress distribution in the composite floor cross-section.

Another alternative is to replace OSB sheathing boards with concrete slabs (reinforced or non-reinforced), as schematically shown in Figure 16a. In this case, the bending load-bearing capacity can be increased if concrete slabs are correctly connected to timber joists. Only a prefabricated floor element behaves as a composite beam element, and can increase the vertical resistance of the whole element. A concrete slab enlarges the static height of an element, and additionally takes the main part of compressive stresses in the element, as schematically shown in Figure 16b. Using the existing (or completely new) timber floor, we can develop an efficient composite system made of timber members in the tensile zone, a concrete layer in the compression zone, and a timber-concrete connection between both elements. Owing to the negative impact humidity has on the strength of timber, it is of utmost importance to install hydro-isolation in the flat area between concrete slabs and timber beams to prevent humidity invasion from the former to the latter (Žegarac and Premrov 2019). Moreover, the minimum depth of a reinforced concrete slab is concreted onto timber beams with the installed steel dowels by simply laying a timber beam with steel dowels onto the concrete plate.

In addition to greater bending resistance, better sound insulation and fire resistance can be achieved with such composite floor elements that can be of the utmost importance especially for timber buildings. Moreover, higher in-plane stiffness of the floor diaphragm can be achieved to improve the seismic behaviour of such buildings (Unuk et al. 2019).

PREFABRICATED TIMBER-GLASS STRUCTURES

The use of glazing in buildings has always contributed to openness, visual comfort, and better daylight situation. Although characterized by weak thermal properties in the past, glass has been gaining an ever-increasing importance as a building material due to its improved thermal, optical, and strength properties, resulting from years of development. The already presented features of wood as a raw, natural, high-ecological building material with an excellent CO_2 footprint, and glass as a highly transparent material with a great possibility to transfer the sun heat through its areas can lead to development of a new type of structures, i.e., so called timber-glass buildings (Figure 17). Envelope building components developed in this way can be suitable for the construction of energy-efficient buildings, in which an the optimal proportion and appropriate orientation of glazing areas play an important role, since they exploit solar radiation as a source of renewable energy within the passive use of energy for heating. With a suitable dynamic form of a building, it is possible to achieve optimal heat demand all year round considering both the heating and the cooling period (Žegarac Leskovar and Premrov 2013, Premrov et al. 2018a). For instance, a suitable or optimal placement of large glazing areas on building envelope elements, which are primary placed on the south façade of a building, enables better energy performance of the building. The daily solar gains obtained through glazing can be evidently higher than the transmission losses throughout the night. More details about energy flows in such cases can be found for example in (Žegarac Leskovar and Premrov 2013), and in many other publications discussing energy behaviour in different climate conditions, and therefore, will not be presented here.

However, such a non-symmetric form of the structure with most transparent areas placed only on one envelope side of the building can lead to a very comfortable energy design. On the other hand, it can be very problematic from the structural point of view when a building is horizontally loaded (i.e., wind and earthquake). If timber-glass wall elements are not load-bearing bracing elements, the rest of the walls (external and internal) without any openings should be able to transmit horizontal load actions acting on the rest of non-transparent wall elements to the basement. In such cases, one of the possibilities is to insert visible diagonal steel or timber elements as main bracing elements. An alternative solution is to insert additional load-bearing internal walls. However, usually none of them is well accepted by architects and investors. Horizontal stability of such timber building can thus become very problematic with increasing insensitivity, especially for multi-story buildings located on very windy or seismic areas.

Figure 17. Dynamic form of a one-family timber-framed house with the primary orientation of glazing areas to the south.

In most cases, the more problematic of these two horizontal load actions is an earthquake, which subjects a building to a high-intensity dynamic load often resulting in catastrophic consequences. One of the basic principles when designing a building to resist seismic loads is trying to avoid plan irregularity, clearly described in, and prescribed by, the Eurocode 8 (2005) standard. This means that the centre of mass (*M*) and centre of stiffness (*R*) of a building should be as close to each other as possible. Unfortunately, this is an issue of energy efficient buildings with large glazing areas predominantly placed on one side of the structure, resulting in an uneven stiffness of their floor plan, and an important dislocation between the centre of gravity and centre of rigidity, as schematically shown in Figure 18.

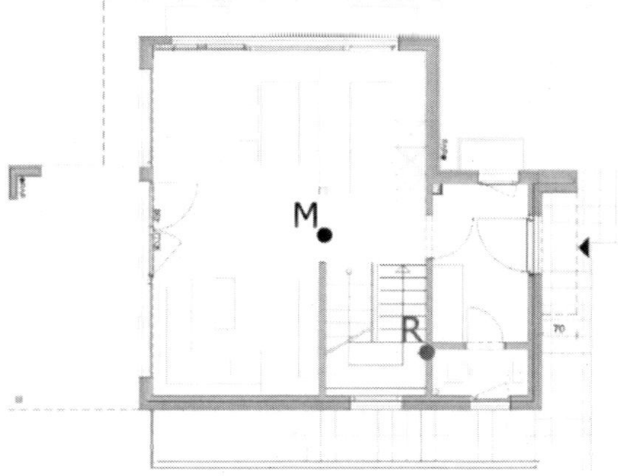

Figure 18. Horizontal distortion of a building due to antisymmetric position of resisting wall elements.

Method A in Eurocode 5 (2005) defines that wall panels which contain a door or window opening should not be considered a contribution to the racking load-carrying capacity. Method B is less restrictive, and states that the lengths of the panel on each side of the opening formed in the panel should be considered as separate panels. Nevertheless, the use of the most accurate numerical FEM approach reveals that wall panels which contain a door or window opening decrease the racking resistance and significantly lower the horizontal stiffness of prefabricated frame-panel wall elements on the one hand, but can still contribute to the horizontal stability of the entire wall assembly on the other. Therefore, the contribution of glass areas to the stiffness of timber-framed wall elements has been rather neglected so far, which can be accounted for by a relatively low strength and ductility levels of ordinary glass used for windows. However, the introduction of modern glass materials, such as tempered and laminated glass or glass-fibre reinforced polymers (GFRP) along with the improvements of glass products' strength properties, has allowed for the use of large glazing areas, since they now contribute to the horizontal stiffness and resistance of wall elements.

To avoid the presented distortion, it is important to consider most external walls on the south facade load-bearing elements. This means that walls with fixed glazing areas (but not windows) should be treated as racking resisting bracing elements and will be treated as composite elements composed of a timber frame and a glass sheathing, which will be somehow able to transmit a considerable part of horizontal forces to the basement. With such an approach, the racking resistance and stiffness of the whole analysed building can be essentially increased, resulting in fewer resisting internal walls or external diagonals to be used to ensure the horizontal stability of the whole building.

Prefabricated Timber-Glass Wall Elements

The demand for the use of glass in timber architecture has been increasing. Several studies have been performed over the past decade to investigate the possibilities of using glass for load bearing elements, for instance (Niedermaier 2003, Cruz and Pequeno 2008, Hochhauser 2011). The main idea of using fixed glazing in prefabricated timber-frame panel wall elements is to replace classical sheathing boards with glass panes, as shown in Figure 19b. Therefore, the research project WoodWisdom (Premrov et. al. 2014) aims at the development of new applications of timber-glass elements, which could provide a certain level of structural resistance, reducing the effect of torsion on wall elements, and at the same time, increase the racking resistance of such timber-glass wall elements. In this case, the racking resistance of the whole analysed building can be essentially increased, and the whole building (Figure 19c) can be statically treated as a composition of classical timber-framed wall elements with OSB or FPB boards (Figure 19a) and so-called timber-glass wall elements (Figure 19b), where the classical sheathing is replaced with a glass pane which should be able to transfer the horizontal load impact to the basement.

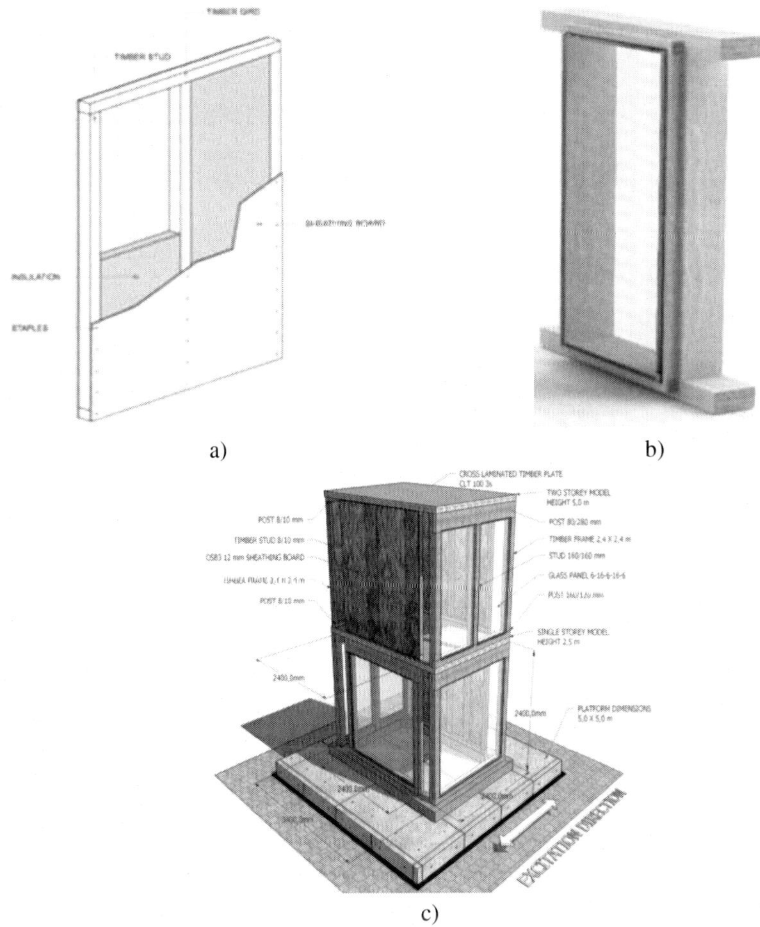

Source: Premrov et al., 2018b.

Figure 19. (a) Timber-framed wall with a classical sheathing board; (b) Timber-glass resisting wall element; (c) Timber-glass box-house model.

The primary aim of analysing timber-glass wall elements was their resistance to a heavy horizontal load impact. The research was performed in three consecutive steps:

- The experimental study of racking resistance on real-scale test samples with insulating three-layered glazing.
- The mathematical modelling of previously experimentally investigated test samples with the approximate mathematical models with axial and shear springs.
- The parametric numerical analysis of the influence of different parameters on the racking resistance and stiffness of timber-glass wall elements.

The experimental study was performed on two different test samples groups (TGWE-1 and TGWE-2) with the insulating three-layered annealed glass directly bonded to the timber frame. The tested timber-glass wall elements (TGWE) consisted of a timber frame with the outside edges measuring 2.4 x 2.4 m. TGWE-1 is a wall element with one large insulating three-layer glass pane in one piece. TGWE-2 is a wall element of the same dimensions, but with two smaller glass panes divided by an additional stud in the middle (Figure 20). The 5.0 mm thick polyurethane adhesive with the end-joint type according to Niedermayer classification was used.

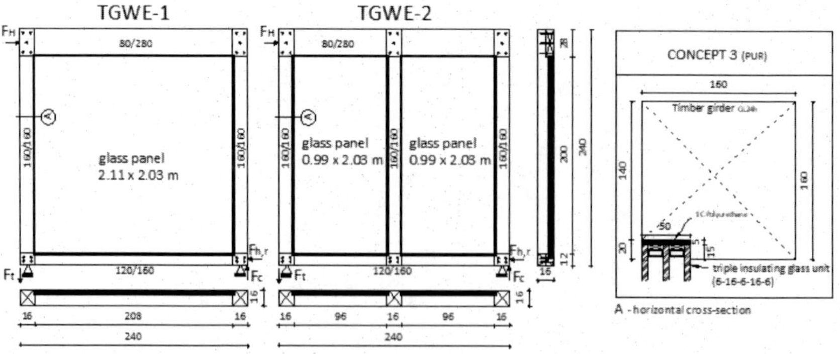

Figure 20. Geometry of the specimens from the testing groups.

Besides the horizontal point loading (*F*) according to EN 594:2011, an additional vertical load of *q = 25 kN/m* was applied onto the specimens to simulate the dead load of the roof and upper floors. The test samples were first subjected to the monotonic horizontal point load (the results are shown in Figure 23) and after that to the cyclic behaviour of the point load. Maximal horizontal displacement (*w*) was measured at the top of the specimens. Further detailed information about the test specimens and the whole testing procedure, and the results can be found in WP 6 final project report (Premrov et al. 2014). The results of the cyclic tests for the test samples of both testing groups are shown in Figure 21.

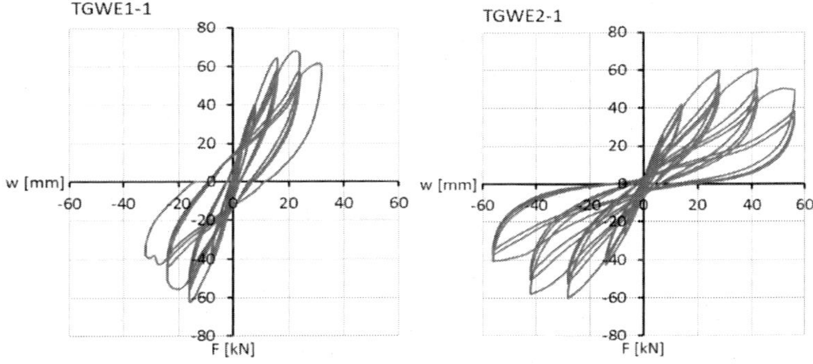

Figure 21. Results of the cyclic test for TGWE-1 and TGWE-2 testing groups.

It is evident that the maximal measured horizontal point load and therefore the racking resistance of the wall elements in the TGWE-1 group are higher than in the TGWE-2 group by 11.7 per cent. The inclination of the hysteresis demonstrates that the racking stiffness of TGWE-1 samples is essentially higher than of TGWE-2 samples. On the contrary, the ductility of the TGWE-2 group is essentially higher than of the TGWE-1 testing group. This is extremely important for the seismic resistance of buildings erected with such timber-glass wall elements, which will be further analysed in the next subchapter.

The mathematical modelling of the previously experimentally investigated timber-glass wall elements was developed according to the so-called spring model presented by (Hochauser 2011). In this model, the flexibility of the bonding line is simulated with the axial and shear stiffness of the bonding line (Figure 22a) to simulate the adhesive joint of a glass-steel connection. The behaviour of a wall element is composed of linear elastic behaviour, followed by nonlinear plastic behaviour (Figure 23b). Initially, a shear flow along the circumference of the glass is present, but when the adhesive starts yielding, a tension field and a compression diagonal are formed, as schematically sjown in Figure a,b. Further details about the calculation of the axial and shear stiffness of the bonding line can be found in (Ber et al. 2018).

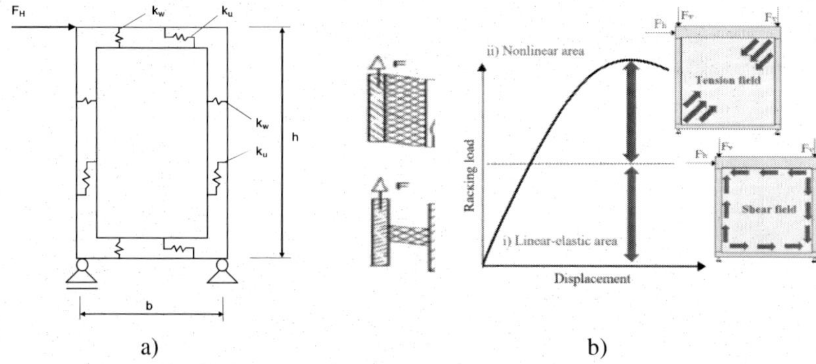

Figure 22. (a) Spring model with an axial and shear spring of a timber-glass wall proposed by (Hochhauser 2011); (b) Response of a timber-glass wall subjected to horizontal racking load in tension and shear behaviour.

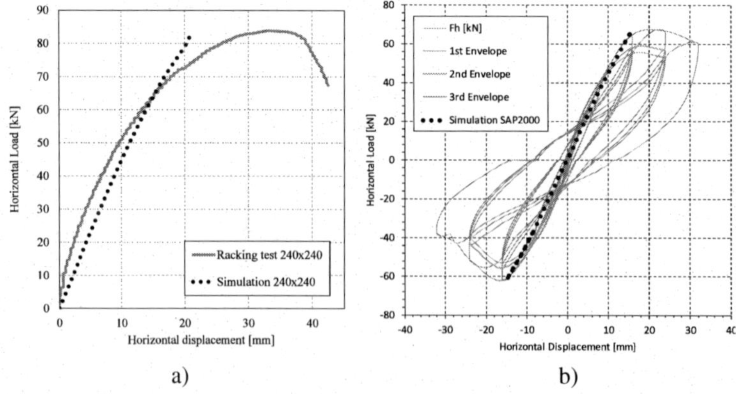

Figure 23. Comparison between the experimental and numerical results of monotonic (a) and cyclic tests (b).

The numerical results obtained with monotonic tests compared to the measured ones for TGWE-1-1 are presented in Figure 23a. It is evident that the numerical results are in a very good agreement with the experimental results in the elastic field, but not so good in ductile behaviour. A comparison of cyclic point load behaviour is shown in Figure 23b and is in a very good agreement with the first envelope of the measured values.

An in-depthparametric numerical analysis of the influence of different parameters on the racking resistance and stiffness of timber-glass wall elements can also be found in (Ber et al. 2018). Unfortunately, it is not presented for three-layered insulating glazing TGWE wall elements, but for our first experimentally tested single-layer glass panes thoroughly presented in (Žegarac Leskovar and Premrov 2013, Premrov et al. 2014, Štrukelj et al. 2015).

Prefabricated Timber-Glass Box-House Models

Combining "classical" prefabricated timber-framed wall elements (Figure 19a) with previously presented prefabricated load-bearing transparent timber-glass wall elements (Figure 19b), it is now possible to construct the so-called timber-glass hybrid building (Figure 19c). Box-house timber glass house models were tested on the shaking table at the IZIIS Institute in Skopje. Four single-story and four two-story box-house structural models (Figure 24) which combine different types of wall elements with ground plane dimension of 2.4 x 3.4 m were tested. Single-story setups had a total height of 2.5 m, while two-story setups reached exactly 5 m in height. Cross-laminated timber floor slabs with 100 mm thickness were used to simulate the rigid behaviour of floor diaphragms and an additional mass of 1,600 kg was installed on each floor to simulate dead load and 30 per cent of the vertical live load in real structural design.

According to the loading protocol, the testing series was divided into two basic stages:

- Low-intensity testing during which the structure remained undamaged and in an elastic state of the material behaviour (including all connections).
- High-intensity testing during which the ground acceleration was scaled up to cause failure in the structure. Before and after each earthquake simulation, a sine sweep test (frequencies in the range of 1–32 Hz, acceleration intensity of 0.01 g) was performed in order to clearly calculate the vibration period of the structure.

Source: Premrov et al. 2018b.

Figure 24. Photo of one and two-story box-house test samples.

Before and after each earthquake simulation, a sine sweep test with frequencies in the range of 1–32 Hz and intensity of 0.01g was performed in order to clearly calculate the natural frequency of the structure, and to record the response of the building. It can be also used to detect damage range to the structure. A diagrams of first periods are shown in Figure 25. The first periods of each test sample were measured before and after the sweep test. If the difference in the first periods is very small, it means that there are practically no deformations in any component of the building. With the increasing of the first period, the stiffness of the building decreases, which results from the loss of rigidity of structural elements (in timber-glass wall elements, this can be the deformation of the adhesive, timber corner or glass), or in the plasticity of metal connecting elements between the prefabricated wall and floor elements. It is expected that the differences in the first periods will be greater for two-story than for single-story buildings.

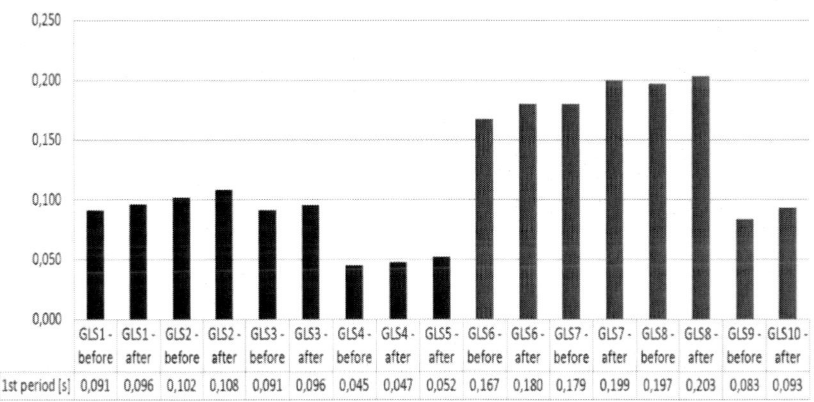

Source: Premrov et al., 2018b.

Figure 25. Diagram of first periods for each model before and after the earthquake simulation.

It may be observed from the presented results that there is only a slight change in the measured first periods before and after excitation by all tested models. This proves that there was only a small decrease in the horizontal stiffness of the tested models, which can be the first indicator that the deformation range in structural elements and connections was not essentially high. In all test samples, no visible deformations in glass elements were detected. The tested walls demonstrated a desirable rocking-type of behaviour without any residual deformations in the adhesive joint, glass panes, and the timber frame. A ductile failure mechanism was established in the steel hold-downs, which satisfies quite a huge part of the ductility of the whole box-house model.

CONCLUSION

As a natural raw material, timber shows indisputable environmental excellence, and is certainly one of the best choices for sustainable construction. Selecting a type of the timber structural system depends primarily on architectural demands, with the orientation, location, and the purpose of a building being of no lesser importance. Most used timber structural systems differ from each other in the appearance of the structure, and in the approach to planning and designing a particular system.

The recent development of new glass products, both in terms of thermal insulation and load-bearing capacity, has enabled completely new architectural aspects in prefabricated timber construction. According to many research studies from the standpoint of energy efficiency, natural lighting and living comfort, the largest area of glazing in buildings must be primarly orientated southward (for buildings in the northern hemisphere), (Zegarac Leskovar and Premrov 2013). Such a position of transparent vertical areas enables better energy performance of a building, where the daily solar gains obtained through glazing can be evidently higher than the transmission losses throughout the night. However, they can cause many structural problems if transparent wall elements are not considered racking resisting vertical elements. In this way and through many different studies mentioned in this chapter, so-called timber-glass wall elements were developed with a quite high share of the horizontal load-bearing capacity. Such transparent wall elements as additional resisting vertical bracings provide designers with many new opportunities to construct prefabricated frame-panel timber buildings with an extremely attractive building shape.

In the last subchapter, certain alternative specialy developed connecting types between timber frames and glass panes are presented. However, there are still many challenges that lie ahead to improve the type of the bonding line, the type of the used adhesive, the position of glazing etc. to enlarge the horizontal resistance and stiffness of timber-glass wall elements, improving the structural stability of the whole timber building. This fact will be especially important in the construction of multi-story prefabricated timber-framed buildings with enlarged glazing placed in the assymetrical position and located in a heavy windy or seismic area, which is one of the most important aspects in the future of sustainable building design in urban areas.

REFERENCES

Abrahamsen, R. B. (2014). The world's tallest timber building. *Tallinn Wood Architecture Conference*, November 2014.

Augustin, M. (2008). Wood Based Panels. In*: Handbook 1 – Timber Structures*. Leonardo da Vinci Pilot Project CZ/06/B/F/PP/168007, Educational Materials for Designing and Testing of Timber Structures.

Ber, B., Premrov, M. and Štrukelj, A. (2016). Finite element analysis of timber-glass walls. *Glass structures & engineering* 1: 19-37. doi:10.1007/s40940-016-0015-4.

Ber, B., Finžgar, G., Premrov, M. and Štrukelj, A. (2018). On parameters affecting the racking stiffness of timber-glass walls. *Glass structures & engineering*: 1-14. doi: 10.1007/s40940-018-0086-5.

Cruz, P. and Pequeno, J. (2008). Timber-Glass Composite Structural Panels: Experimental Studies & Architectural Applications. *Conference on Architectural and Structural Applications of Glass*. Delft University of Technology, Faculty of Architecture, Delf, Netherlands.

European Committee for Standardization CEN/TC 250/SC5 N173. (2002). *EN 1991-1-1: Eurocode 1: Actions on structures - Part 1-1: General actions - Densities, self-weight, imposed loads for buildings*, Brussels.

European Committe for Standardization CEN/TC 250/SC5 N173. (2005). *EN 1995-1-1:2005 Eurocode 5: Design of Timber Structures, Part 1-1 General rules and rules for buildings*, Brussels.

European Committee for Standardization. (2005). *EN 1998-1 Eurocode 8: Design of structures for earthquake resistance – Part 1: General rules, seismic actions and rules for buildings*, Brussels.

European Committee for Standardization. (2011). *EN 594:2011: Timber structures – Test methods – Racking strength and stiffness of timber frame wall panels*, Brussels.

Gold, S and Rubik, F. (2009). Consumer attitudes towards timber as a construction material and towards timber frame houses – selected findings of a representative survey among the German population. *Journal of Cleaner Production* 17: 303–309.

Hochhauser, W. (2011). *Ein Beitrag zur Berechnung und Bemessung von geklebten und geklotzten Holz-Glas-Verbundscheiben* [*A Contribution to the Calculation and Measurement of Adhered and Clipped Wood-Glass Components*]. PhD diss., Vienna University of Technology.

Kolb, J. (2008). *Systems in Timber Engineering*. Birkhäuser Verlag AG, Basel - Boston – Berlin.

Niedermaier, P. (2003). Shear-Strength of Glass Panel Elements in Combination with Timber Frame Constructions. *Proceedings of the 8th International Conference on Architectural and Automotive Glass (GPD)*, 262-264, Tampere, Finland.

Premrov, M. and Kuhta, M. (2009). Influence of fasteners disposition on behavior of timber-framed walls with single fibre-plaster sheathing boards. *Construction and Building Materials* 23(7):2688-93. doi:10.1016/j.conbuildmat.2008.12.010.

Premrov, M. and Kuhta, M. (2010). *Experimental Analysis on Behaviour of Timber-Framed Walls with Different Types of Sheathing Boards*. Construction Materials and Engineering, Nova Science Publishers.

Premrov, M. and Dobrila, P. (2012a). Numerical analysis of sheathing boards influence on racking resistance of timber-frame walls. *Advances in Engineering Software* 45(1): 21-27. doi:10.1016/j.advengsoft. 2011.09.012.

Premrov, M. and Dobrila, P. (2012b). Experimental analysis of timber-concrete composite beam strengthened with carbon fibres. *Construction and Building Materials* 37: 499-506. doi:10.1016/ j.conbuildmat.2012.08.005.

Premrov, M., Serrano, E., Winter, W., Fadai, A., Nicklisch, F., Dujič, B, Šušteršič, I., Brank, B., Štrukelj, A., Držečnik, M., Buyuktaskin, H. A., Erol, G. and Ber, B. (2014). *Workshop report "WP 6: Testing on life-size specimen components: shear walls, beams and columns including long-term behaviour"*. Woodwisdom-net, research project, load bearning timber-glass-composites, 2012-2014.

Premrov, M., Žigart, M. and Žegarac Leskovar, V. (2018a). Influence of the building shape on the energy performance of timber-glass buildings located in warm climatic regions. *Energy* 149: 496-504. doi:10.1016/j.energy.2018.02.074.

Premrov, M., Žegarac Leskovar, V., Ber, B. and Štrukelj, A. (2018b). Advantages of energy-efficient timber-glass buildings. *Proceeding: 1st International Conference on Sustainability, Technology and Busi*ness, Melaka, Malaysia, (ICSTB 2018).

Štrukelj, A., Ber, B. and Premrov, M. (2015). Racking resistance of timber-glass wall elements using different types of adhesives. *Construction & building materials* 93: 130-143. doi:10.1016/ j.conbuildmat.2015.05.112.

Unuk, Ž., Premrov, M. and Žegarac Leskovar, V. (2019). Development of an innovative approach for the renovation of timber floors with the application of CLT panels and structural glass strips. *International journal of architectural heritage: conservation, analysis and restoration*, [Online ed.], Published online: 15 Jul 2019: 1-17, doi:10.1080/15583058.2019.1637479.

Vogrinec, K., and Premrov, M. (2013). Mathematical modelling of timber-framed walls using fictive diagonal elements. *Applied mathematical modelling* 37(16/17): 8051-8059, doi:10.1016/j.apm.2013.02.050.

Waugh, A., Weiss, K. H. and Wells, M. (2009). *Process Revealed Auf dem Holzweg* [*Process Revealed On The Wood Path*]. London: Murray & Sorrell FUEL.

Žegarac Leskovar, V. and Premrov, M. (2013). *Energy-Efficient Timber-Glass Houses*, Springer. London, Heidelberg, New York, Dordrecht: Springer.

Žegarac Leskovar, V. and Premrov, M. (2019). *Integrative Approach to Comprehensive Building Renovations*, Green Energy and Technology, Cham, Switzerland: Springer International Publishing.

About the Author

Miroslav Premrov, PhD

University Professor

University of Maribor, Faculty of Civil Engineering, Transportation Engineering and Architecture (UM FGPA), Maribor, Slovenia

Miroslav Premrov is the author or co-author of 52 research papers indexed by the Science Citation Index, and a reviewer for the most reputable international journals from the field of civil engineering structures. As a member of WG 5, he is an actively participates in the preparation of Slovenian standards in timber structures. In 2013, he published a scientific monograph "Energy-Efficient Timber-Glass Houses" through Springer Verlag, which was selected by Slovenian research agency as the best scientific contribution in civil engineering in Slovenia that year. As a co-author, he published a scientific monograph "Integrative approach to comprehensive building renovations" through Springer Verlag in 2019. Since 1999, he has been actively dealing with the problems of strengthening composite timber elements used in multi-story prefabricated timber buildings. In 2010, he commenced a wide research in timber-glass buildings, considering load-bearing capacity problems integrated with energy-efficient concepts of such structures. The main findings are published in three international scientific monographs and in many journals. Between 2012 and 2016, he was the head of the Slovenian research team in the international FP 7 Wood Wisdom project entitled "Load-bearing timber-glass composites", and at the same time, he was also the head of Work Package 6 "Testing on life-size specimen components" of this international research project.

In: An Interdisciplinary Approach … ISBN: 978-1-53617-302-4
Editors: V. S. Klemenčič et al.

Chapter 6

APPLICATION OF INTERDISCIPLINARY APPROACH: "PUNKT PODLEHNIK" WORKSHOP

***Ivana Banfić*[*], *Tina Belec*[*], *Tadej Božak*[*], *Jovana Dragović*[*], *Klara Glad*[*], *Sebastijan Kelc*[*], *Jani Knuplež*[*], *David Leskovar*[*], *Katja Nežmah*[*], *Nikola Pintarić*[*], *Selma Rogač*[*], *Simon Sekereš*[*], *Sanja Špindler*[*], *Gregor Tollazzi*[*], *Ivana Vuković*[*], *Miroslav Premrov*[#], *Marko Renčelj*[#], *Vanja Skalicky Klemenčič*[#,†], *Vesna Žegarac Leskovar*[#] and *Maja Žigart*[#]**

University of Maribor, Faculty of Civil Engineering, Transportation Engineering and Architecture, Maribor, Slovenia

---

[*] Students; # Mentors

[†] Corresponding Author's Email: vanja.skalicky@um.si.

ABSTRACT

The interdisciplinary student workshop entitled “Punkt Podlehnik” selected for the course “Sustainable concepts of building design” in the 2017/18 winter semester addressed the challenge of sustainable development of the Municipality of Podlehnik. On the following pages, a student project addressing the central residential area with additional public programmes in Podlehnik is presented as a graphical appendix. Students have been guided by interdisciplinary mentors to examine design challenges from social, environmental, and economic perspectives. Various phases of sustainable building design are shown, such as a site visit and an analysis of the location, the conceptual design phase and the project development phase.

Keywords: application, interdisciplinary, student workshop

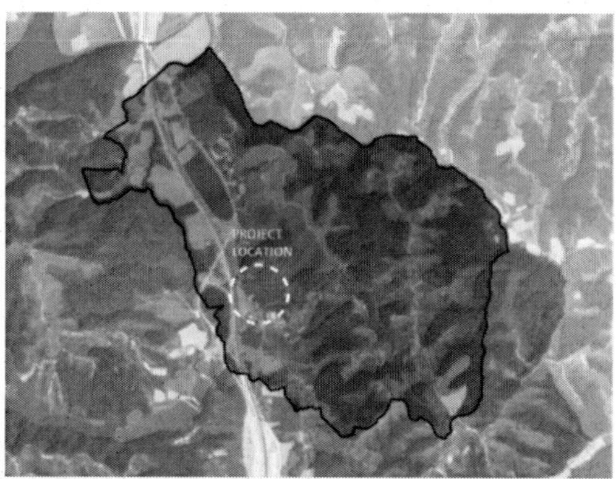

Figure 1. Project location.

Figure 2. Students and mentors’ site visit to the municipality.

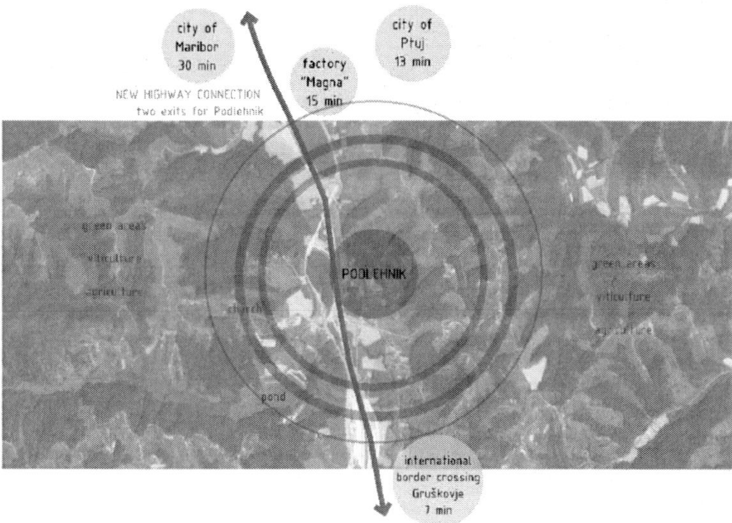

Figure 3. Analysis | Connectivity.

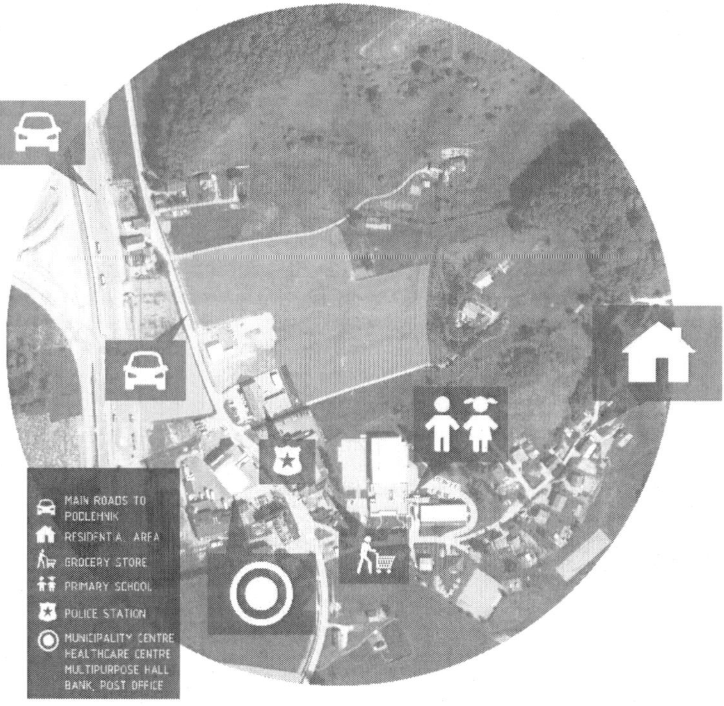

Figure 4. Analysis | Existing programme.

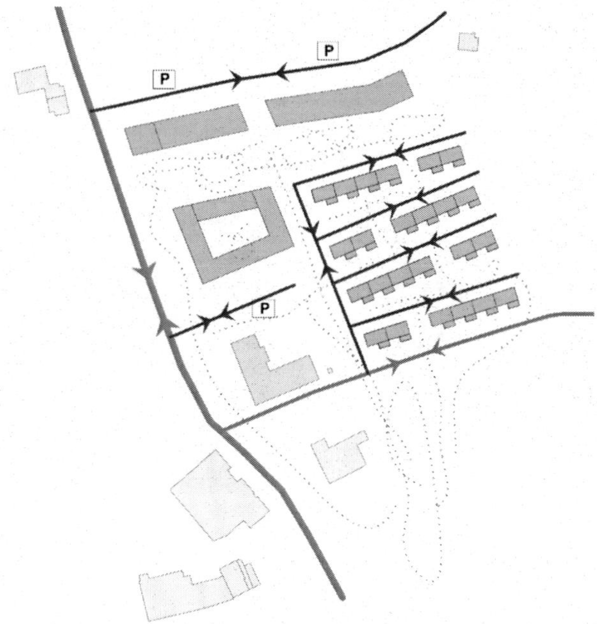

Figure 5. Concept | Traffic design.

Figure 6. Concept | Green areas design.

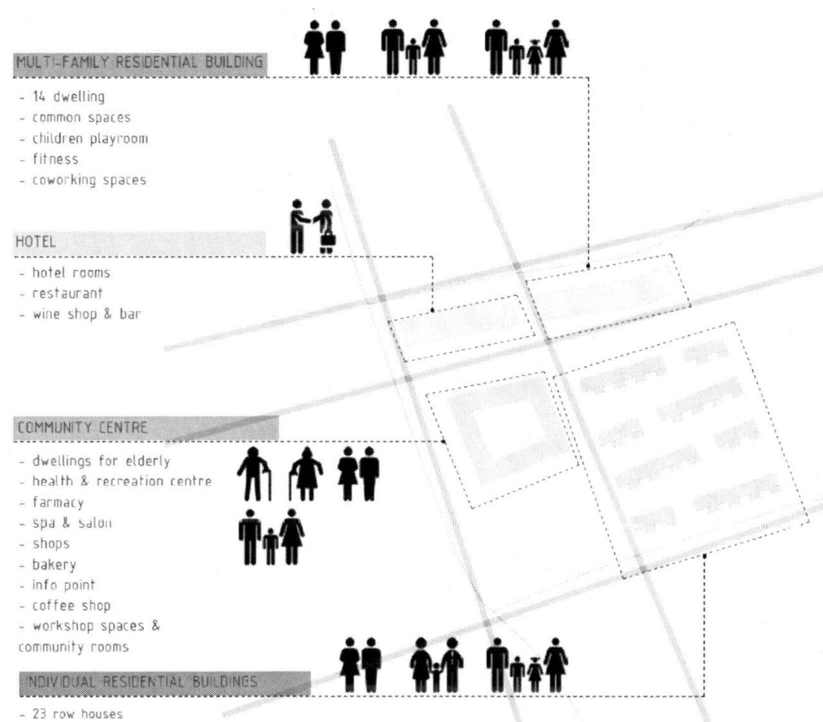

Figure 7. Concept | Programme design.

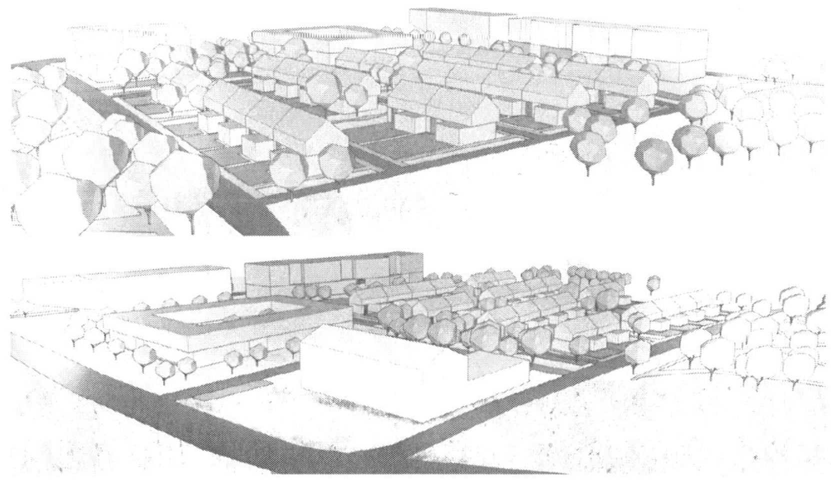

Figure 8. Situation | Schematic 3d.

Figure 9. Situation | Plan.

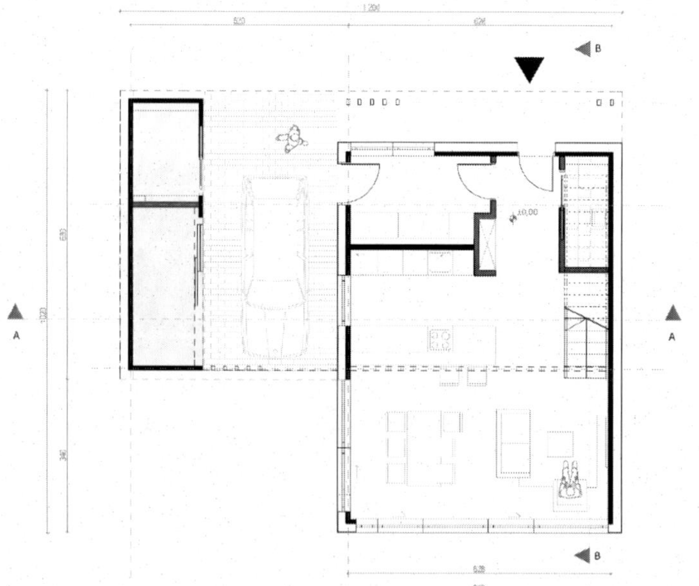

Figure 10. Row house type A | Ground floor plan.

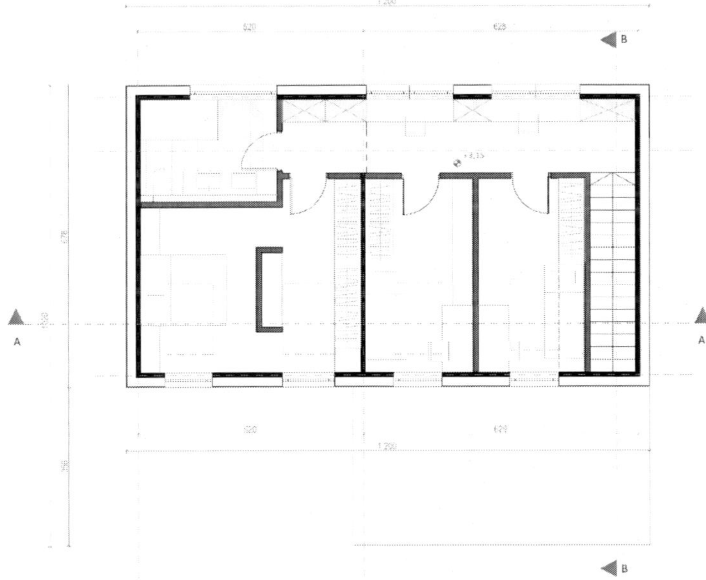

Figure 11. Row house type A | First floor plan.

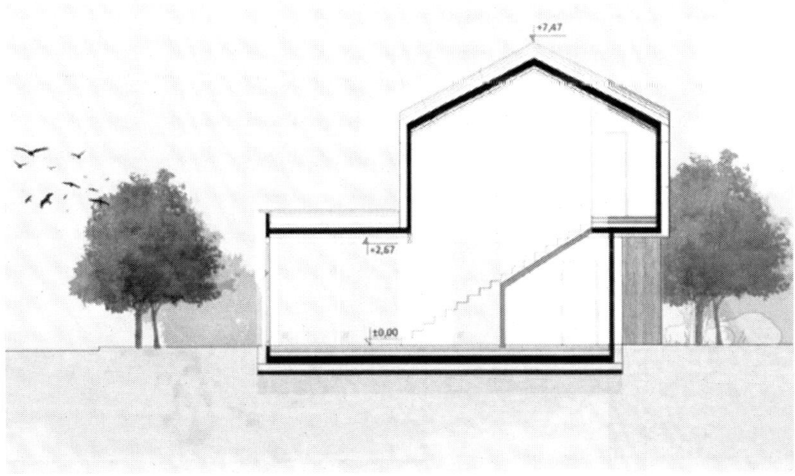

Figure 12. Row house type A | Section A-A.

Figure 13. Row house type A | South façade.

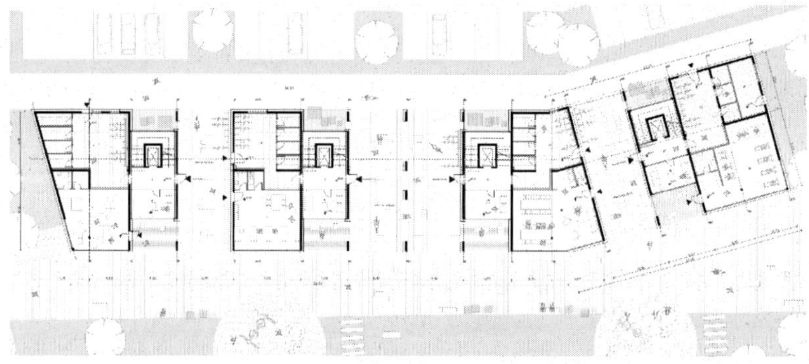

Figure 14. Multi-family residential building | Ground floor plan.

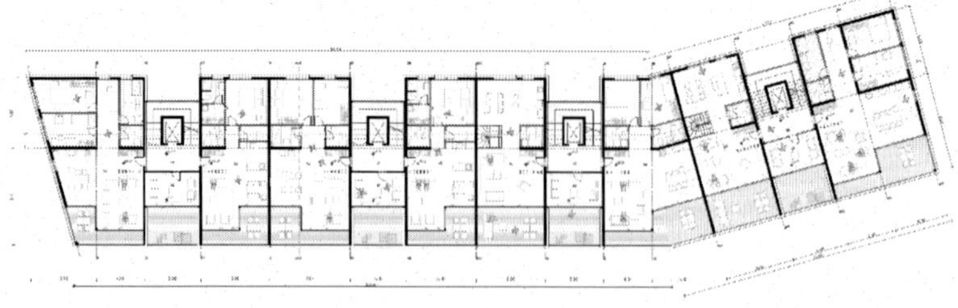

Figure 15. Multi-family residential building | First floor plan.

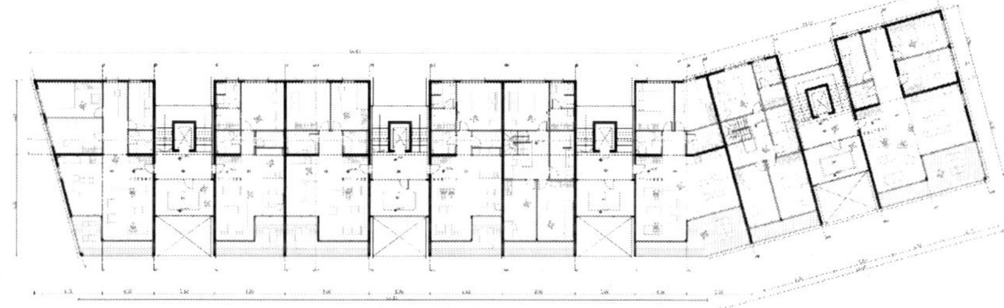

Figure 16. Multi-family residential building | Second floor plan.

Figure 17. Multi-family residential building | Cross section and east façade.

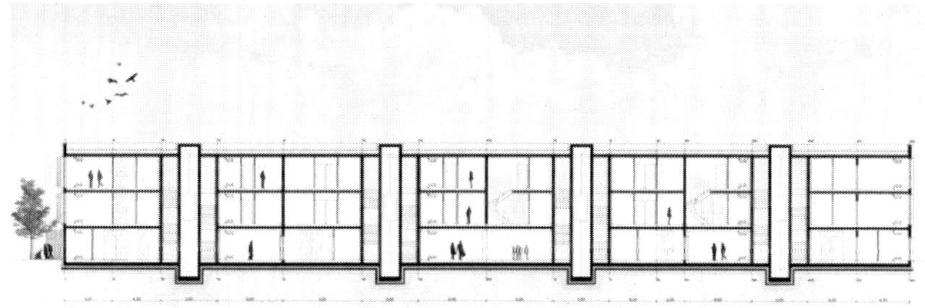

Figure 18. Multi-family residential building | Cross section.

Figure 19. Multi-family residential building | South façade.

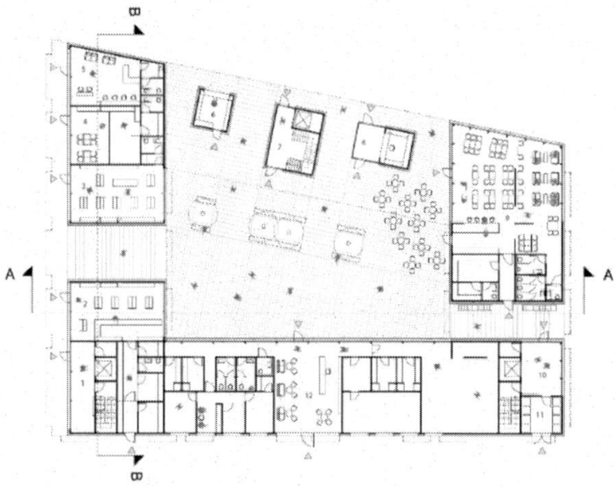

Figure 20. Podlehnik Community Centre | Ground floor plan.

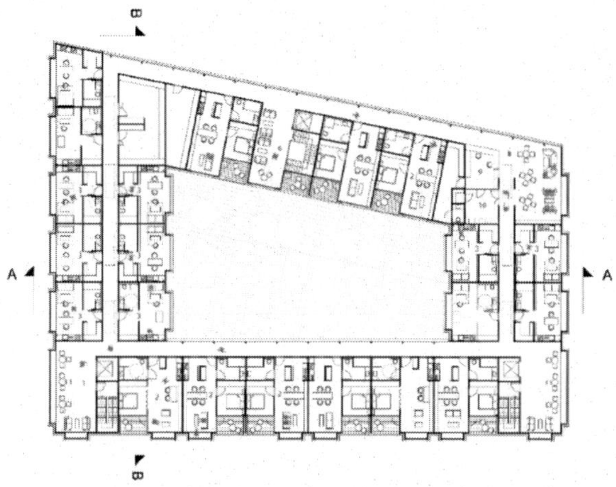

Figure 21. Podlehnik Community Centre | First floor plan.

Figure 22. Podlehnik Community Centre | Section A-A.

Figure 23. Podlehnik Community Centre | East façade.

Figure 24. Podlehnik Community Centre | South façade.

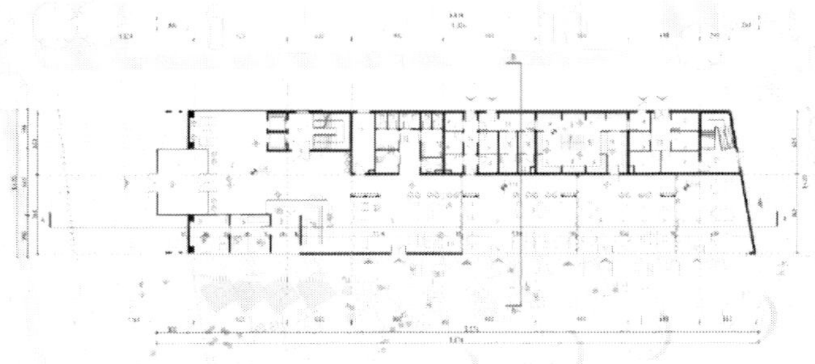

Figure 25. Hotel | Ground floor plan.

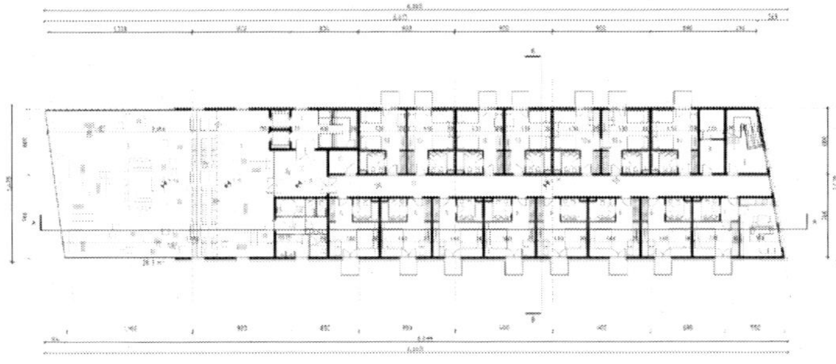

Figure 26. Hotel | First floor plan.

Figure 27. Hotel | Section A-A.

Figure 28. Hotel | South façade.

Figure 29. Hotel | North façade.

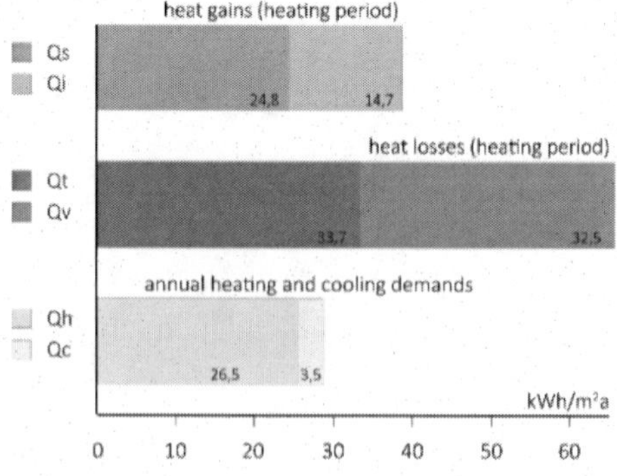

Figure 30. Example of energy performance results | Multi-family residential building.

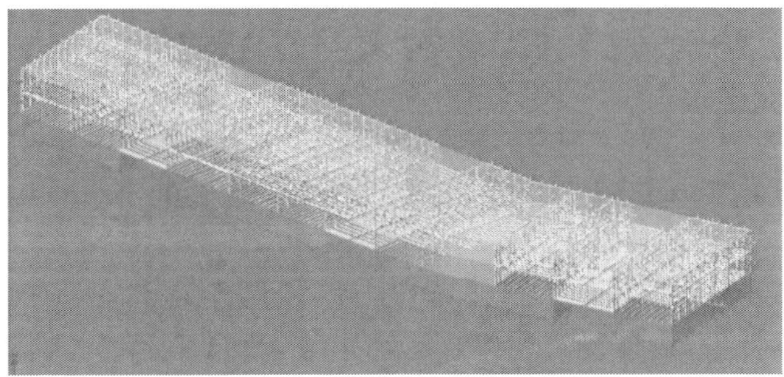

Figure 31. Example of 3d model of construction | Multi-family residential building.

Figure 32. Row house | Visualization.

Figure 33. Multi-family residential building | Visualization.

Figure 34. Podlehnik Community Centre | Visualization.

Figure 35. Hotel | Visualization.

About the Editors

Vanja Skalicky Klemenčič, PhD
University Teaching Assistant
University of Maribor, Faculty of Civil Engineering, Transportation Engineering and Architecture (UM FGPA), Maribor, Slovenia
Email: vanja.skalicky@um.si

Vanja Skalicky Klemenčič graduated from the Faculty of Architecture at the University of Ljubljana, where she finished her doctoral thesis with the title "Contemporary Scandinavian Urban Planning Principles and Criteria for High-Quality Living in Residential Environments". Her PhD research also took place at the Norwegian University of Life Sciences in Ås. The main purpose of the visit was to learn about the Scandinavian interdisciplinary and integrated approach in residential design. Her research is focused on different aspects of comprehensive urban and residential development, and designing from environmental and social sustainability to comprehensive high-quality urban design. She works as a research and teaching assistant for architecture and spatial planning at the Department of Architecture of the Faculty of Civil Engineering, Transportation Engineering and Architecture, University of Maribor. She has been involved in various international research projects as a co-mentor in numerous urban

– architectural workshops, as well as a co-author of different sustainable urban design development projects and competitions.

Vesna Žegarac Leskovar, PhD
University Associate Professor
University of Maribor, Faculty of Civil Engineering, Transportation Engineering and Architecture (UM FGPA), Maribor, Slovenia
Email: vesna.zegarac@um.si

Vesna Žegarac Leskovar is an associate professor of architecture and spatial planning at the University of Maribor, Faculty of Civil Engineering, Transportation Engineering and Architecture (UM FGPA). After initially educating and working as an architect, she received a Ph.D. in Architecture at the Technical University Graz, Faculty of Architecture. At UM FGPA, she teaches courses on sustainable architecture, architectural structures, and technology, energy-efficient building design with a particular focus on timber buildings. She is the leader and a team member of numerous research and development projects carried out by UM FGPA for national and European industrial partners and research organisations. Additionally, she supervises diverse students' workshops which deal with the topic of sustainability in architecture. Her research focuses on the issues of sustainable building design, and covers the area of energy efficiency, bioclimatic design, environmental assessment of buildings, and building renovation. In addition to being the author of many international conference contributions and publications, she publishes her research in international journals indexed by JCR and SNIP. She is the co-author of two scientific monographs published by Springer Verlag. The first one, "Energy-Efficient Timber-Glass Houses", was released in 2013 and the second one, "Integrative Approach to Comprehensive Building Renovations", in 2019. The first monograph was selected by the Slovenian Research Agency (ARRS) as the best scientific contribution in civil engineering in Slovenia in 2013.

Maja Žigart
University Teaching Assistant
University of Maribor, Faculty of Civil Engineering, Transportation Engineering and Architecture (UM FGPA), Maribor, Slovenia
Email: maja.zigart@um.si

Maja Žigart has graduated in architecture from the Faculty of Civil Engineering, University of Maribor, in 2013. During her studies, she actively engaged in the study process at the faculty as a student tutor, and participated in two student workshops. She has been working as a teaching assistant for architecture and spatial planning at the Faculty of Civil Engineering, Transportation Engineering and Architecture at University of Maribor since 2014. Her areas of expertise are architectural design and visualisation, energy-efficient timber buildings, and sustainable design of buildings. At the faculty, she engages in many student workshops and research projects which involve sustainable design of buildings. Her research has been presented at conferences and published in international journals. She enrolled in the doctoral programme at the Institute of Buildings and Energy at the Technical University of Graz in 2016.

INDEX

#

A

B

C

D

E

F

G

H

I

L

M

N

O

P

Q

R

S